CULTURE, PLACE, AND NATURE
Studies in Anthropology and Environment
K. Sivaramakrishnan, Series Editor

Centered in anthropology, the Culture, Place, and Nature series encompasses new interdisciplinary social science research on environmental issues, focusing on the intersection of culture, ecology, and politics in global, national, and local contexts. Contributors to the series view environmental knowledge and issues from the multiple and often conflicting perspectives of various cultural systems.

Mountains of Blame

CLIMATE AND CULPABILITY
IN THE PHILIPPINE UPLANDS

Will Smith

UNIVERSITY OF WASHINGTON PRESS
Seattle

Mountains of Blame was made possible in part by grants from the Samuel and Althea Stroum Endowed Book Fund, the Australian Academy of the Humanities, and the Alfred Deakin Institute at Deakin University.

Copyright © 2020 by the University of Washington Press

Composed in Warnock Pro, typeface designed by Robert Slimbach
Cover design by Katrina Noble
Cover photograph: Barangay Tagabinet, Palawan, Philippines, 2017. Photograph by Jason Houston for USAID.
Interior map and photographs are by the author.

24 23 22 21 20   5 4 3 2 1

Printed and bound in the United States of America

All rights reserved. No part of this publication may be reproduced or transmitted in any form or by any means, electronic or mechanical, including photocopy, recording, or any information storage or retrieval system, without permission in writing from the publisher.

UNIVERSITY OF WASHINGTON PRESS
uwapress.uw.edu

LIBRARY OF CONGRESS CATALOGING-IN-PUBLICATION DATA
LC record available at https://lccn.loc.gov/2020020429
LC ebook record available at https://lccn.loc.gov/2020020430

ISBN 978-0-295-74815-3 (hardcover), ISBN 978-0-295-74816-0 (paperback), ISBN 978-0-295-74817-7 (ebook)

Portions of chapter 2 were previously published in "Rooted in Place? The Coproduction of Knowledge and Space in Agroforestry Assemblages," *Annals of the American Association of Geographers* (© 2017), reprinted by permission of Taylor and Francis Ltd. (www.tandfonline.com). Portions of chapter 3 were previously printed in "Governing Vulnerability: The Biopolitics of Conservation and Climate in Upland Southeast Asia," *Political Geography* (© 2019), reprinted by permission of Elsevier. Chapter 4 was previously published as "Weather from Incest: The Politics of Indigenous Climate Change Knowledge on Palawan Island, the Philippines," *Australian Journal of Anthropology* (© 2018), reprinted by permission of John Wiley and Sons.

The paper used in this publication is acid free and meets the minimum requirements of American National Standard for Information Sciences—Permanence of Paper for Printed Library Materials, ANSI Z39.48–1984.∞

CONTENTS

Foreword by K. Sivaramakrishnan   vii

Acknowledgments   xi

List of Abbreviations   xiii

Introduction   3

CHAPTER ONE
Making *Uma*, Imagining *Kaingin*   31

CHAPTER TWO
Rooted in Place   61

CHAPTER THREE
Insidious Vulnerabilities   93

CHAPTER FOUR
El Niño and Incest   111

Conclusion: Placing Blame   129

Glossary   139

Notes   141

Bibliography   153

Index   167

FOREWORD

The literature on climate change emerging from sociocultural anthropology has begun to move beyond tired debates on mitigation and adaptation, coping with disasters, and the creation of climate refugees who are displaced and driven from homes and livelihoods. Along the way it has become increasingly nuanced in considering historical, moral, and politically inequitable processes through which climate disruption is experienced, explained, and countered. In this vein, Will Smith has produced a study that is attentive to longer histories of indigenous people in the forested uplands of Palawan Island, a part of Mindanao in the Philippines. The region and its residents are no strangers to the climate variations produced by El Niño currents and the storms that beset the South China Sea. But in the last few decades, they have observed new weather patterns that are perceived as linked to the social transformation they have undergone in reorganizing their agriculture and immediate relations to plants, soil, and society.

With empathy and care, Smith examines the self-awareness and critical evaluations of their own actions and responses by the indigenous people of Palawan. The injustice of the way in which native people have borne the brunt of climate change–related disasters—and before that, the rapacity of extractive industries in their territories—has generated studies critiquing such ravages in emic indigenous terms (see, e.g., Povinelli, 2016; De la Cadena, 2015), even as they analyze the limitations of liberal social theory in synthetic, often sweeping, arcs (e.g., Tsing, 2017; and a proliferating literature on the Anthropocene). Smith moves beyond analyses that fail to differentiate between experiences of environmental risk and disruption across societies and their particular histories. Such histories, while learning to be more attentive to the place and role of nonhumans, remain distinguishable by local cultural processes and world-connecting forces that emerge from human action and sensibility.

In that respect, this work is more akin to that of Tania Li (2014) and Karine Gagné (2019) in its patient and respectful attention to the way local perspectives on larger environmental forces and economic processes are dynamic, steeped in self-appraisal, and generative of commentary on both wider conditions and personal foibles as a product of the changes that have enmeshed human lives. Without losing sight of marginalization and degradation as a macrosocial and powerful set of inexorable tendencies in their lands and lives, indigenous Pala'wan seem also to rue what they have become. They reflect thereby on a moral crisis that crept up on them along with the resource exploitation and bad weather that has tormented and impoverished them. Like the Diné people of Navajo Nation described by Dana Powell (2018), the Pala'wan are aware of both colonial and postcolonial injustice and their own sad alienation from sources of internal community renewal and strength.

In this work, Smith integrates perspectives on space, place, and power, on the one hand, with concerns with affect, ontology, and nonhuman life and its variable sentience on the other hand. His commentary on the plural meanings of the Tagalog and Pala'wan terms for livelihood provide good examples of the ways in which indigenous ontologies and political commentaries are generated simultaneously. As the study moves from an analysis of swidden farming in the forested uplands to the increased volatility of weather and attendant perspectives on climate variability, it remains grounded in concerns with indigenous knowledge: its production and transmission, under conditions of rapid socioecological changes in a locale challenged by economic, environmental, and political stress.

Both conservation and resource development, as Smith notes, are products of exercises in assigning blame and then enacting countermeasures that transform people and landscapes into objects for invasive or protective measures. This leads Smith to propose and use, quite effectively, the notion of postfrontier. He is referring in part to the continued sporadic nature of state attention to the southern districts of Palawan, where the forests remain home to indigenous, largely rural communities, even as the urban settlements nearby are dominated by Christian and Muslim migrants. After pondering original interpretations of El Niño–influenced drought and subsequent distress and changes in cultural norms and social behavior that are put forward by the people in the forests of Palawan, he then returns to the powerful themes of moral assessment and commentary on calamity and suffering, not in fatalist terms but to parse the personal from the global in chains of consequent action that might be of significance.

K. SIVARAMAKRISHNAN
Yale University

ACKNOWLEDGMENTS

Like all research produced through ethnographic fieldwork, this book has accrued some substantial, and complicated, forms of personal and intellectual debt. My thanks, first and foremost, must go to the Pala'wan people of Bataraza and, in particular, those who suffered my lengthy and tedious interviews focusing on the minutiae of *uma* production. I am grateful for the hospitality of many upland households that hosted me during my stays, but in particular the families that regularly shared their rice and bamboo floors with me. I am also indebted to the Pastor, who was a constant companion and source of support during my many misadventures in the forests of Inogbong. His relatives welcomed me into their home and supported much of my research by acting as impromptu research assistants. I must also thank the many other government employees and migrant household members in Bataraza who made time to talk with me over the course of my fieldwork.

Like many Palawan scholars, I have been fortunate to enjoy the support of my spouse-colleague, Sarah Webb, who has shared in my fieldwork on Palawan and many other trips to the Philippines. She provided essential critique and commentary over the course of the research and writing process. This book would simply not exist without her. Our daughter, Wattle, in her own way, provided feedback and support for the project. My fieldwork on Palawan Island has also benefited tremendously from a fortuitous overlap with Noah Theriault, who has indulged me over the years in fine-grained discussions of environment and indigeneity in southern Palawan. I am further indebted to the generosity of many Filipino friends and colleagues on Palawan and beyond—Ched Limsa, Dr. Carin Alejandria, and Dr. Ramon Docto, among many others—who shared their time and insights with me. I am also grateful for Raul Pertierra for his hospitality while in Manila.

Various parts of manuscript have benefited from the comments and feedback of colleagues, supervisors, mentors, and anonymous peer reviewers too numerous to mention individually. I owe a special thanks, however, to Wolfram Dressler, whose infectious enthusiasm for the study of swidden agriculture has left an indelible intellectual mark on both this book and myself.

The writing up of this book was financially supported by a fellowship funded by the Bushfires and Natural Hazards Cooperative Research Centre and the Alfred Deakin Institute for Citizenship and Globalisation at Deakin University. I am grateful for the advice from many of my colleagues at ADI, but in particular the mentorship of Tim Neale, who provided some expert guidance on the publishing process. I also appreciate the detailed commentary on the manuscript and guidance on the publishing process provided by Lorri Hagman at the University of Washington Press, in addition to two anonymous reviewers who provided generous and thoughtful critique.

ABBREVIATIONS

CBFM	community-based forest management
DENR	Department of Environment and Natural Resources
ECAN	Environmentally Critical Areas Network
ENSO	El Niño–Southern Oscillation
FAO	Food and Agriculture Organization
IP	Indigenous People
IPRA	Indigenous Peoples Rights Act
MMPL	Mount Mantalingahan Protected Landscape
NGO	nongovernmental organization
PAMB	Protected Area Management Board
PCSD	Palawan Council for Sustainable Development
PIADP	Palawan Integrated Development Project
PTFPP	Palawan Tropical Forestry Protection Programme

MOUNTAINS OF BLAME

Introduction

In social movements and protests unfolding around the world, both climate change and the vulnerability of indigenous people are positioned as the product of violent dispossession and ongoing industrialization. Indigenous peoples, it is argued, who have borne the largest burdens of structural violence emanating from colonial and neoliberal processes and contributed the least to greenhouse gas emissions, suffer unfairly at the front lines of environmental transformation. Blame for anthropogenic climate change, and therefore responsibility to act and ameliorate harm, lies with wealthy nations whose prosperity is derived from the unchecked burning of fossil fuels and decades of environmental degradation. At the same time that indigenous people occupy a position of absolute innocence, their intimate knowledge of the environment and ecologically harmonious lifestyles offer potential salvation in the form of adaptation and mitigation strategies. As the role of indigenous peoples in climate change narratives suggests, motivating changes in global environmental policy is dependent on leveraging these vast and compelling disparities in culpability between polluters and poor, often rural, peoples. But what does it mean when indigenous people say, and act, otherwise?

On Palawan Island, an environmental postfrontier located in the southwest of the Philippine archipelago, indigenous Pala'wan people in the forested uplands of the municipality of Bataraza have confronted the conceptual and material realities of changing weather. As elsewhere in Southeast Asia, recurring El Niño and La Niña events have been amplified by anthropogenic climate change and now increasingly disrupt life for indigenous households throughout Palawan Island. However, rather than placing culpability for change on Western industrialization, many indigenous people on Palawan Island have arrived at conclusions and put into practice strategies that confound the moral expectations of international policy makers, researchers, and climate activists: Pala'wan often blame themselves for the climatic uncertainty that regularly impacts indigenous lives and livelihoods. When discussing changes in the weather, many Pala'wan people observe that whereas in the past the onset of the rainy season was fixed, typically arriving in March or April, monsoonal rainfall patterns have become increasingly unpredictable. Before, the weather was moderate and provided an ideal balance of rain and sunshine. Now, floods and droughts regularly devastate agricultural production. Customary leaders, in particular, suggest that these climate changes are a divine punishment for the increase of *sumbang*, or incestuous relationships, occurring in the area. Rather than rallying to "traditional" and perhaps "well-adapted" agricultural practices, Pala'wan men and women instead actively seek out precarious waged labor in lowland fields and monocropped paddy rice or cash crops and, in doing so, seemingly render *themselves* increasingly vulnerable to cyclical El Niño droughts.

Pala'wan accounts of climate change, and others that confront notions of scientific causality and the moral calculus of established anthropogenic climate change narratives, are both ethically and conceptually unsettling. The past twenty years have seen an explosion of scholarly literature from a variety of disciplines concerned with understanding how indigenous peoples are impacted by, and respond to, anthropogenic climate change.[1] This work paints a collective portrait of peoples who both have an intimate knowledge of their environment and are experiencing, through no fault of their own, disruption resulting from unpredictable weather conditions. Indigenous peoples not only feature prominently in discussions over the social and economic ramifications of a changing climate but also are now identified in global environmental policy as a vital source of knowledge, practices, and lifeways that can help others adapt to and mitigate the impact of dramatically changing weather patterns. This indigenous victim-savior complex is both widespread and powerful. Prior to the institutionalization of indigeneity in global governance frameworks in the 1990s, accounts of climate change that disrupt the moral and instrumental logic of established environmental narratives could have been easily relegated to the margins of nongovernmental organization reporting, state policy, and ethnographic accounts (if not excised completely from the scholarly record). However, the growing role of indigenous people within debates over the origins of and responses to climate change has forced these discomforting views onto the attention of scholars, activists, and policy makers who often find themselves asking: How and why do those who have contributed least to the emission of greenhouse gases and environmental degradation come to position themselves as culpable for the devastating impacts of climate-related disasters?

This question defies easy explanation or totalizing theories of human-environmental relationships. Much of recent efforts to confront these complexities is focused on providing a kind of moral resolution to unsettling accounts of environmental change through the work of contextualization. Expanding on the efforts of anthropologist Mary Douglas and others to politicize blame and "danger" (Douglas, 1992; Giddens, 1999), critical environmental scholars[2] have worked to rehabilitate accounts of troubling blame by placing these narratives within histories of colonial domination and neoliberal exploitation. This form of analysis might unveil what anthropologist Peter Rudiak-Gould (2013, p. 142) has identified as the hidden "counter-hegemonic" potential of unsettling climate change narratives. In Rudiak-Gould's ethnographic work on the Marshall Islands, he argues that Marshallese narratives of self-blame for anthropogenic climate change do not represent ignorance or "false consciousness" but can be interpreted as a critique of a perceived cultural decline driven, ultimately, not by the islanders themselves but by a seductive and environmentally destructive modernity. By carefully shifting the locus of critique from the self to wider and more powerful forces, as Rudiak-Gould does, scholars reframe morally distressing explanations of marginalized people for climate change or other forms of environmental rupture so that they are "not as disturbing as [they] may first appear" (Rudiak-Gould, 2013, p. 142). This approach draws attention to the often highly power-laden nature of weather in human societies, and a wide body of historical research has examined the relationship between control of weather and precolonial and colonial systems of domination, hierarchy, and extraction.[3] This framing often considers local climate discourse among agrarian households to be a "manifestation of debates over the efficiency, legitimacy and morality of social arrangements" (Sheridan, 2016, p. 237). Ethnographically contextualizing such accounts offers the potential to reconfigure unsettling cultural difference into metaphoric parables of uneven power relations strategically deployed by the marginalized, achieving what anthropologist Elizabeth Povinelli (2002, p. 107) has referred to as the "ethnographic magic" of "transcoding ... moral horrors into reasoned/reasonable difference."

While this book focuses on historically and ethnographically contextualizing environmental narratives in complex fields of power, it is less concerned with bending alternative accounts of climate change to fully resolve the moral tension arising from self-blame. Such efforts do not represent a universal analytical strategy for resolving troubling accounts of environmental change and can provide ethically palatable explanations only in the context of very specific political and economic histories. Instead, I focus here on the generative potential that emerges from probing unsettling (and perhaps unresolvable) ethical narratives in relation to the ways in which indigenous peoples are understood and leveraged in global climate change scholarship and policy. Rather than excising troublesome explanations for climate change or seemingly self-induced vulnerability from the academic record or reconfiguring them into acceptable stories, what have been referred to as "flat narratives of innocence" (Hughes, 2013), I argue that they present an opportunity to interrogate and revise key ideas that underpin global environmental governance and development interventions that depend on, target, and impact the lives of indigenous peoples who reside within ecologically sensitive and valued landscapes. Through ethnographically exploring the experiences of Pala'wan households that deviate from established climate change narratives in which indigenous peoples are unambiguously victims, this book troubles dominant conceptualizations of vulnerability, questions the way indigenous knowledge is utilized within international climate change discourse, and seeks more broadly to unsettle issues of blame and culpability in the study of human-environmental interaction.

Pala'wan accounts of climate change exist in relation to perennial conflicts and struggles over how to manage tropical forests, and for whose benefit this management does and should occur. In Southeast Asia, "the uplands," or mountainous and often still-forested hinterlands, are not just fixed geographic entities but the product of power-laden social and economic relations over thousands of years that have shaped forest ecologies, ethnic identities, and patterns of livelihood. Since the turn of the twentieth century, many indigenous people on the Philippine island of Palawan have discovered that they now reside within state-owned forestlands. These forestlands are of intense interest to people both in the Philippines and throughout the world as dwindling reserves of biodiversity, sources of economic wealth, stores of carbon, and sites of increasingly commodified ecosystem services. Because they are located in these rare and valuable spaces, indigenous peoples and their activities have become subject to intense scrutiny and evaluation and, in many cases, interventions aimed at redeeming what are often perceived as their environmentally destructive use of forests. The Philippine uplands in which Pala'wan and other indigenous peoples respond to environmental change can be considered "mountains of blame": spaces where interlocking concerns over culpability for environmental decline and developmental concerns over the backwardness of ethnic minorities converge and come to inform the everyday lives of upland peoples. Self-blaming narratives of indigenous peoples can be understood only as positioned within, and informed by, these wider moral ecologies of environmental discourse and material intervention into upland lives and livelihoods. *Mountains of Blame* explores how indigenous peoples on Palawan Island must contend with changing weather in the context of shifting efforts to govern and transform their way of life. Self-blaming narratives occur within a historical project of marginalization that has come to inform everyday livelihood practices, traumatic experiences of hunger and deprivation, and the production of environmental knowledge by indigenous Pala'wan households. Self-blame is entangled with the ways relations between people and place are envisioned and managed.

PUTTING WEATHER IN PLACE

Over the past two decades, anthropologists engaging with political ecology have increasingly turned to ideas of space in the study of conservation and development. These ethnographic engagements with the spatiality of governance schemes and resource management projects have explored what might be called struggles over place. More than the representational politics of human-environmental interaction, I draw on these perspectives to focus on the routine moral logics that underlie and justify interventions into lives and livelihoods and the everyday material practices that constitute processes of belonging and place making. These approaches emphasize how processes are connected over time and are shaped by uneven relations of power that link seemingly peripheral places to global practices of imagining and influencing nature. Rather than isolated and preexisting, "places, in short, are constructed historically in processes that spatially exceed the local and in which the extralocal is as constitutive as the local" (Biersack, 2006, p. 16).

However, while relations between sovereign states, nongovernmental organizations, and rural communities may always be asymmetrical in the coproduction of spatialities, environmental anthropologists concerned with the everyday nature of power and place have demonstrated the partial and fragmented nature of efforts to craft conservation landscapes. Donald Moore (2005, p. 12), for example, has argued in his exploration of postcolonial land rights in Zimbabwe that "micropractices matter in grounded geographies. Where cultural practices *take place* matters because they are among the critical assemblages that *produce place*." In Moore's conception of place, the ability to shape landscape is not monopolized by state power but operates at the entanglement of multiple spatialities that are overlain, or "sedimented," over time to produce landscapes (p. 21). Similarly, Arturo Escobar has drawn on the assemblage thinking of philosophers Félix Guattari and Gilles Deleuze to argue for a politics of place in understanding environmental conflict. In *Territories of Difference*, Escobar (2008, p. 67) argues that "the politics of place can be seen as an emergent form of politics, a novel political imaginary in that it asserts a logic of difference and possibility that builds on the multiplicity of actions at the level of everyday life." In seeking to counter state-driven agendas, Escobar describes how black and indigenous activists in Colombia work explicitly to tie culture and environment to notions of territory through both representational and embodied practice. Place, then, becomes "a crucial dimension of the making not only of local and regional worlds, but also of hegemonies and resistance to them" (p. 30).

In Paige West's (2006) seminal examination of conservation landscapes in the Papua New Guinea highlands, she calls attention to the ways in which places are produced at the intersection of uneven power and conflicting understandings of nature. Informed by the work of geographer Henri Lefebvre, she argues that places "are not a given, not locations that come into being with ecology and evolution, but rather are produced by the social and material relations between people" (p. 28). Despite their marginal position, Gimi villagers mediate the work of conservation by drawing biologists, project officers, and other outside actors into a reciprocal economy of social obligation. In surveying the often-disparate aims of villagers and conservation practitioners, West asks, "Are Gimi producing a landscape in one way and the biologists in another given their histories and subjectivity? Is there a hybrid production of nature going on at Crater?" (p. 31). Through projects of environmental governance, such as the efforts to establish a protected area at Crater Mountain, various spatial productions do not unfold in parallel but are instead "made and folded into each other" (p. 32).

West's work recognizes the power of local categories of analysis to explore the multithreaded constitution of space. How are cultural constructs discursively and materially deployed, and how do they inform multiple spatialities of conservation landscapes? To understand how conservation schemes unfold and transform people and places in the Philippines, I focus on notions of "livelihood"—a term long used by development scholars and practitioners to conceptualize the productive practices of rural peoples—to think about how struggles over place are entangled with experiences of weather in complex and unexpected ways. The dominance of livelihood as a conceptual frame in recent rural scholarship has its origins in household studies literature, and through the 1980s became a central feature of scholarship on participatory forms of development and conservation practice, in large part through the work of rural sociologist Robert Chambers. Over time, livelihood has been the subject of significant conceptual elaboration. Successive iterations of "livelihood frameworks" have been deployed as a practical tool that can more accurately capture and catalog productive capabilities of rural households in both scholarly analysis and development schemes.[4]

However, rather than draw on livelihood as a conceptual scheme that simply renders diverse local practices legible to scholarly analysis (and amenable to governmental intervention), the notion of livelihood has considerable purchase as a shared conceptual category for both state agencies and rural peoples. In Hannah Bulloch's (2017) recent ethnography on the Philippine island of Siquijor, she argues that the seemingly universal cross-cultural appeal of "development," deployed freely by both politicians in Manila and the rural poor, masks the intensely local and personal reworkings of progress under this umbrella term. Livelihood is similarly a common grammar between those who govern and the governed themselves. The term is an interface through which diverse and often-conflicting meanings of people and nature are constructed, communicated, and put into practice by both indigenous Pala'wan and Philippine forest policy. "Indigenous livelihoods" and their spatial implications are the representational and material terrain through which conflicting ideas of environment come to produce "place" in much of the Philippines. The densely humanized forest landscapes in the Philippines are constituted through how people in rural areas produce food and seek cash income, what they consume, and where they seek to live in relation to these activities.

Critiques of inclusive conservation practices that nominally involve local resource users in formal decision making have often focused on the socially constructed nature of targeted "communities" (Brosius et al., 2005). While simplifications of community also pervade Philippine environmental governance, it is the concept of a "livelihood" that has become an essential yet poorly examined "boundary object" for forest management. Through a focus on livelihoods, policy makers and planners aim to replace what are deemed to be environmentally destructive ways of being and living in the landscape (primarily swidden agriculture, but also varied forms of nontimber forest product collection, hunting, and other "tribal" practices) with what are often ambiguously termed "alternative livelihoods." Such alternative possibilities for living envisioned by the state are diverse, but often reflect shifting visions of economic and environmental modernity. They are, typically, market based and intensely faddish in their prescriptions of particular crop or livestock options, often incorporating prevailing trends in regional or national development practice. In the process of establishing and undertaking forest management, the desires and needs of indigenous households are imagined, and alternatives are devised, which are then provided and promoted through projects of environmental governance. The management of livelihood entails not only the provision of specific kinds of tools, seeds, and other materials but also a comprehensive reorganization of indigenous lives and subjectivities in order to produce the intended conservation spaces. Since the colonial period, the livelihoods of indigenous people have slowly but steadily become a central focus of environmental politics in the Philippines, the "object-target" of government, in Michel Foucault's terminology (1978, p. 166), in which new ways of being and living in forest landscapes must be taught by state agents and learned by rural households.

These efforts speak to the work of critical environmental scholars who draw on Foucault's notion of governmentality to examine the production of environmental subjectivities in forested landscapes in developing nations.[5] These conceptualizations emphasize the significance of power-knowledge relationships in environmental regulation as the means of indirectly shaping environmental behavior, or what Foucault has famously termed the "conduct of conduct." More recent appraisals of Foucault's work have come to see governmentality beyond the emphasis on conditioning power/knowledge dyads that has dominated much of this literature. For the geographer Kevin Grove (2013), this has meant a turn to Foucault's notion of biopolitics to understand the multiplicity of environmental politics. Rather signaling a specific kind of governmental modality or set of techniques of management that are immutably associated with particular times or regimes, for Grove, "biopolitics signals a problem space in which life is made amenable to calculated programs of governmental intervention and improvement" (p. 27). This perspective provides a more expansive set of possibilities that can accommodate the institutional messiness and confusion through which forests and livelihoods are actually managed in the Philippines, where forest governance has increasingly focused over time to manage not just "environmental" practices or knowledge but the intimate features of indigenous lives in the service of conservation: rather than singular and monolithic governmental "rationality," conservation efforts may be composed of fractured or even competing efforts that mutate, accrete, or fall away over time. This means that disciplinary measures, regulation of "life itself" in the exercise of biopower, and the conditioning of power-knowledge relationships may all work together to shape subjectivities in unexpected ways with unintended effects.

Of course, governmental interventions do not perfectly remodel subjectivities and discipline unruly human bodies to reorder them in space. Forest peoples hold their own visions of economic and household futures that may converge or diverge from the imaginaries of planners and policy makers—often in ways beyond the plans and intentions of grand conservation schemes. A focus on livelihood as the assembled and everyday "tasks of making a living" can attend to how environmental rule is tempered and resisted in place (Ingold, 2000, p. 5). In anthropologist Juno Perreñas's (2018) recent ethnography of orangutan rehabilitation in Malaysia, she argues that the environmental subjectivities of indigenous Iban workers are informed not only by conservation practices but also by wider experiences of seeking and making a living. She draws on the Malay idiom *cari makan* (to find food) to argue that forms of displacement and loss experienced in making a living come to powerfully inform environmental subjectivities and actions.

Drawing on these perspectives, which emphasize the embodied elements of making a living, my use of livelihood in relation to Pala'wan productive or consumptive practices here does not refer to an immutable or cohesive cultural ideal, but orients analysis towards emic perspectives, everyday politics, and deep histories that lie behind such tasks and their spatiality. On Palawan Island, the Tagalog term for livelihood, *kabuhayan*, or the Pala'wan term *pengkebiyagan* is sometimes used by indigenous Pala'wan people to articulate a range of meanings. Either can refer variously to a source of food, a self-conscious reflection on tradition, aspirations for cash income, or an unchanging condition of poverty. In an oral history conducted early in my fieldwork, I asked an older indigenous man how the livelihoods (*kabuhayan*) of indigenous people have changed since he was small. Despite what I had understood to be dramatic social and ecological changes that had occurred on Palawan Island, especially since the 1950s, his response was, "Nothing has changed. It's just the same." Clearly, there is a playfulness and strategy in how the language and practice of specific livelihoods are deployed, often in ways that both resist and co-opt state conservation agendas. Livelihoods on Palawan Island are therefore focal points where wider resource politics become entangled and contested through the everyday actions of households and communities targeted for reformation. Focusing in such detail on indigenous livelihood practices can therefore point to the shifting and unstable intermingling of state and household agencies that can account for, as West describes, how multiple spatialities may be "folded" into one another.

In this book, I interweave struggles over place with indigenous Pala'wan experiences of weather. A focus on livelihoods in the first half of this book, as both assemblages of embodied and material practices with spatial effect *and* contested objects of governance and discursive production, links the politics of forest management to concrete matters of experiencing and responding to changes in the weather in the second half. This focus on contested spatial production and its implications for life and livelihood, when brought into conversation with well-established conceptualizations of rural vulnerability and indigenous knowledge of climate change, challenges the simplifications of culpability and blame that pervade discourses of indigeneity and environmental change.

PLACING SWIDDEN AGRICULTURALISTS IN THE UPLANDS OF SOUTHEAST ASIA

In Southeast Asia, questions of forest degradation often focus on the uplands as distinct, though often vaguely bounded, spaces of management. These hilly and mountainous hinterlands are not only environmentally distinct from lowland plains but also viewed as places of cultural difference and marginality that are symbolically and legally inscribed onto the landscape. Yet these spaces are not natural spatial or biogeographic units of analysis. That the uplands are culturally and environment distinct, as places where there are still forests (or places "once forested" or "to be forested again") and ethnic minorities reside, is historically contingent. In Anna Tsing's oft-quoted explanation from *In the Realm of the Diamond Queen* (1993, p. 90), these "out of the way places" of social and environmental difference are not the untouched product of primordial isolation but an "ongoing relationship with power" that stretches back centuries. They are correspondingly sites where conflict over the use of resources tends to be exaggerated and highly racialized, and are often linked to questions of citizenship, development, and nationhood. These "mountains of blame" are intensely moralized spaces where culpability for environmental decline can mobilize significant political and economic resources for the sake of forest conservation—and in doing so come to shape the lives of those who reside there.

Central to anxieties surrounding forest degradation throughout Southeast Asia, perhaps more than any other form livelihood or human activity within these landscapes, has been the role of swidden agriculture. "Swidden"—also known as "shifting cultivation" or more pejoratively as "slash-and-burn agriculture"—is a term used to refer to diverse rotational forms of agricultural production that involve the clearing, drying, and burning of forest cover to sequester nutrients into the soil. Once exhausted, plots are typically abandoned to fallow. Despite the now exotic character attached to the practice, swidden was until recently widespread throughout the world's forests, including parts of Europe and Japan into the 1960s (Sigaut, 1979), and was the predominant form of agricultural production in Southeast Asia prior to European colonization. Today, swidden agriculture is largely confined to mountainous landscapes throughout the tropical world. Despite trajectories of decline in recent decades, it remains a vital source of subsistence and cash income for communities throughout Southeast Asia (van Vliet et al., 2012). However, since the inception of the region's colonial regimes, swidden agriculture has been frequently understood by state foresters as the key driver of tropical forest decline, a threat to the economic potential of colonial governments, and later, an object of intense developmental concern.

Anthropologist Michael Dove (1983, p. 86) has noted that debates surrounding swidden in Southeast Asia deal not with "the reality of this system of agriculture but rather a much distorted, mythical conception of it." What sustains this "ignorance" surrounding shifting cultivation, he suggests, are the political economic interests served by demonizing upland small-holders: the promotion of commercial plantations, the appropriation of lands construed as degraded, large-scale logging operations, and more recently carbon sequestration and other Payment for Ecosystem Services schemes. Control over forested landscapes, in many Southeast Asian nations, is often dependent on reductive representations of swidden cultivators as economically wasteful and environmentally destructive. These myths, in sum, have facilitated and continue to facilitate "the extension of external control into the territories of swidden agriculturalists" (p. 96). These characterizations, I suggest, are an important aspect of discussions surrounding what Nancy Peluso and Peter Vandergeest (2001) have termed the "political forests" of the region, the moral narratives that drive ongoing interventions into indigenous lives and are an important though often ignored element in rural experiences of changing weather. The specific features of shifting cultivation, such as its mobility and reliance on fire, have come to inform narratives of blame and motivate livelihood interventions that shape the lives of upland peoples (figure I.1). These powerful, but also incomplete, interventions constitute upland spaces and increasingly form the political economic and biophysical basis through which indigenous households negotiate climate change on Palawan Island.

The "Austronesian hypothesis" suggests that swidden agriculture became the dominant form of land use in both the coastal plains and the mountainous hinterlands in much of insular Southeast Asia following the expansion of Austronesian people from Taiwan from approximately 3000 BCE

FIG. I.1 An indigenous Tagbanua man burning a swidden plot in central Palawan

(Bellwood, 2006). It was not until the twentieth century that now-iconic irrigated paddy rice farming became the norm for most communities, especially in much of the Indonesian and Philippine archipelagoes, and swidden became an activity relegated to people and places increasingly framed as marginal and backward in national discourses. James Scott (2009) has argued that prior to European colonization the uplands of montane Southeast Asia had historically been composed of peoples retreating into the ungovernable mountains to escape the excesses of lowland states that came to rely on complex and fixed systems of irrigated rice production. Underlying the possibility of these nonstate spaces was, firstly, the friction of mountainous terrain, which inhibited the ability of lowland powers to coerce and reliably extract wealth from upland areas. Further confounding the ability of lowland administrators to render legible human populations for the purposes of management and taxation was the seemingly chaotic nature of swidden livelihoods, which formed the basis of subsistence for most communities living in densely forested areas. Swidden fields were typically highly mobile and multicropped with perishable tubers that could not be easily identified, let alone appropriated, by centralized forms of power. For both these reasons, Scott argues that the mountainous uplands could operate as an "extrastate . . . catchment zone for refugees from state-making projects in the valleys" (p. 31).

This dynamic was profoundly altered for much of the region following colonialization by European powers over the course of the seventeenth to twentieth centuries. Driven by economic imperatives, colonial interest in Southeast Asia was focused on the exploitation of forest resources. Teak, in particular, was sought after as a valuable commodity for shipbuilding, and expansive stands in Myanmar (then Burma) and Indonesia fueled the military incorporation of indigenous polities into wider colonial endeavors. While a wide variety of local forest uses were criminalized, such as the collection of firewood and valuable nontimber forest products or the firing of lands for grazing, swidden agriculture was seen as the largest threat to the economic value of forests.

Colonial environmental narratives throughout the region came to position swidden agriculture as the primary agent of deforestation and in doing so identified various ethnic minorities as wasteful destroyers of forest resources. In Vietnam, swidden agriculture become the "avowed enemy" of the French colonial forest service (McElwee, 2016, p. 37). In narratives of forest decline, blame was readily placed on upland peoples: "The methods of cultivation of the mountain tribes . . . quickly leads to real destruction of the wooded area. . . . Huge areas are ruined piece by piece, and we do not expect this situation will improve, as it has not been possible to settle these wandering tribes" (Service Scientifique de l'Agence Économique de l'Indochine, 1931, p. 2, as cited in McElwee, 2016). Similarly, in colonial Burma under British rule, swidden or *taungya* not only was linked to the inevitable decline of forests and valuable teak stands but also was implicated in growing "soil erosion, flooding and localized drought" (Bryant, 1997, p. 69). Concerns over the environmental impact and economic efficiency of mobile forms of cultivation were often articulated through cultural anxieties focused on the backwardness of particular ethnic groups. In Sarawak, for example, British colonial officials connected the persistence of swidden agriculture and the hesitance of Iban farmers to fully engage in modern economic life, seen here as the burgeoning plantation economy, to their continued residence in communal longhouses and other customary practices (Cramb, 2007, p. 126).

Responses by colonial forestry agencies to these concerns and the impact on the livelihoods of upland peoples were varied. In British Burma, there were efforts to incorporate *taungya* into teak production and financially incentivize sedentary forms of agricultural production as a means of enhancing valuable timber reserves. The American colonial regime in the

Philippines optimistically emphasized the reeducation of recalcitrant upland farmers ("propaganda") as a means to solve deforestation, albeit in recognition of the limited resources available to enforce a blanket ban on already outlawed livelihoods.[6] Broadly, however, swidden was criminalized through fines and threats of imprisonment.

The total elimination of rotational agriculture was a nominal and largely unquestioned goal throughout the region's colonial forestry services. Attendant to punitive forms of management were new forms of zoning and boundary making through which swidden agriculture could be excluded from forests and, ultimately, eliminated as a practice. Colonial governments readily claimed lands as crown or state property that, in turn, required systems of classification aiming to categorize newly obtained areas as either more valuable as timber stands or as released for agricultural development. While initially colonial administrators were stymied by harsh terrain and mobile people, improving communications and military technology and, perhaps more importantly, Western notions of political space and government changed the relationship of upland communities to lowland state power. In Dutch colonial management of Javanese forests, the need to manage forest resources led to the establishment of varied forms of demarcation and visualization (e.g., regulated teak forests and nonteak *wildhoutbosschem*, or "wildwoods"), which could conceptually separate economically valuable forests (Peluso, 1992, pp. 52–53). These governance spaces flattened complex indigenous systems of resource access to criminalize what were seen to be wasteful practices of rural small-holders. These kinds of territorial strategies, which fixed and institutionalized upland-lowland divisions, were widespread in the nineteenth century and dependent on emerging theories and practices of European "scientific forestry" (Vandergeest, 2003). In this way, forestry can be understood as central to the creation of the uplands as legally, ethnically, and ecologically distinct spaces.

The independence of formerly colonial nations during the mid-twentieth century did little to change policies toward the agricultural practices of ethnic minority groups. Throughout much of the region, forestry remained strongly influenced by received wisdoms of the colonial period. Both prior to and after the Second World War, more in-depth research with swidden agricultural communities helped partly revise the dominant simplifications of upland peoples as primitive and wasteful and opened debate over the environmental and economic impacts of swidden. The work of anthropologists such as Derek Freeman (1955) in Borneo, Edmund Leach (1954) in Burma, and George Condominas (1957) in Vietnam provided systematic accounts of an agricultural system previously understood largely as arbitrary and irrational. Most notable was the influential work of anthropologist Harold Conklin (1957), whose extraordinarily detailed exploration of Hanunóo agriculture on the Philippine island of Mindoro emphasized the longstanding cultural significance and "integral" nature of swidden for many communities for whom shifting cultivation was an enduring way of life rather than a temporary phase on an evolutionary ladder. In reflecting on these combined efforts, it was possible for geographer Joseph Earl Spencer (1966, p. 4), who had studied Philippine swidden systems, to tentatively claim in his monograph reviewing shifting cultivation in Southeast Asia that he "would enter the plea that not all aspects of shifting cultivation are destructive, that we must know what it is we seek to improve . . . and that the total replacement of shifting cultivation by sedentary agriculture will bring more ills than it will cure."

However, these insights were not evenly translated into forest policy or broader public opinion, and swidden agriculture remained a key feature of deforestation narratives readily embraced in the postcolonial period. Forestry services largely maintained and expanded existing systems of forest management and punitive approaches to swidden management. During this period, forced relocations were a common means to address converging environmental, social, and military problems that upland peoples often represented. In Indonesia, for example, thousands were targeted for relocation from ancestral lands with only cursory support for new sedentary livelihoods in order to preserve timber for logging interests (Lindayati, 2002). Partly, this can be seen as the product of institutional inertia. European officials were gradually replaced with locally trained staff, which created cohorts of indigenous foresters inculcated with antiswidden narratives and views of forest use focused on economic exploitation. These views were also reinforced by influential international research groups such as the United Nations' Food and Agriculture Organization (FAO). FAO publications at the time would emphasize that "shifting cultivation is the greatest obstacle not only to the immediate increase of agricultural production but also to the conservation of the production potential for the future, in the form of soils and forests" and that swidden "is not only a backward type of agricultural practice. It is also a backward stage of culture in general" (FAO, 1957).

Exclusionary practices and forced relocations became incrementally less popular as the means to manage forest resources following the Second World War. In their place, what became known as "social forestry" initiatives culminated with the wide institutionalization of community-based forest management during the 1990s. In an approach arguably pioneered in Southeast

Asia by foresters in the Philippines, putatively more inclusive social forestry programs sought to draw in local peoples to cooperatively manage forest resources rather than criminalize or exclude them. Despite being seen as a break from the harsh and punitive approach to swidden management of the past, newer community-based interventions continued to be informed by longstanding environmental narratives and biases against mobility and subsistence production. In cynical interpretations, community-based interventions often simply entailed more intensive and sophisticated management of communities rather than the devolution of any meaningful control (Agrawal, 2005). In many cases, social forestry efforts were simply seen as a more effective way to achieve older goals that oriented around the removal of swidden and the economic rationalization of state forests. As such, programs targeting many communities continued to frame swidden as a problem to be eliminated rather than attempt to understand local aspirations or indigenous forms of resource management. This perspective was well summarized by FAO publications that noted in the 1980s of swidden's future in the region: "the only alternative seems to be devising land management systems that reduce the fallow period or eliminate it altogether" (FAO, 1984, p. 169).

Over the same period, swidden management widened beyond simply environmental concerns into larger efforts to govern upland spaces for the purposes of development, and in doing so, it has become increasingly ethnicized in the postcolonial context. In Thailand, for example, swidden has been closely associated with "hill tribes," whose resource use is blamed for deforestation, watershed decline, and localized climate change, meaning that environmental management is frequently bound up with wider interethnic conflicts unfolding across the lowland-upland gradient (Forsyth & Walker, 2008). This association of swidden agriculture with cultural difference and people who have become "ethnic minorities" or "indigenous peoples" means that the practice was caught up in twentieth-century development anxieties over the progress of distinct mountain peoples who were, and often continue to be, targeted on the basis of their supposed backwardness (Tsing, 1993).

This book's focus on the politically inflected nature of swidden management is not to romanticize or essentialize upland people and places. To frame swidden simply as a sustainable practice unjustly maligned by environmental science ignores enormous regional diversity in political-environmental histories and agroecological practices. It also downplays farmers' own responses, many of whom have come to embrace the sedentary and market-based forms of modernity proffered by states and civil society. Instead, this description of swidden agriculture demonstrates the ways in which ideas about people, place, and environment have significant afterlives and can come to shape what are known as upland spaces, local knowledge, and livelihoods in unexpected ways. It is in this context of a wider moral ecology of blame surrounding deforestation and biodiversity loss, and attendant efforts to imagine and govern indigenous peoples' way of life in the uplands, that Pala'wan people are experiencing changing weather. Indigenous people and their livelihood practices have been constructed as the primary agents of deforestation in colonial forest policy, and these representations of people and place endure to shape governmental interventions, and to impact everyday life in complex ways.

INDIGENOUS PEOPLE AND FORESTS ON PALAWAN ISLAND: LIFE ON A PHILIPPINE (POST)FRONTIER

The Philippines, perhaps more than most other nations in the region, is a country consumed by its own environmental degradation and the ecological fate of its uplands (Broad & Cavanagh, 1993). Early European explorers visiting the archipelago marveled at forests stretching, unbroken, from coasts to mountain peaks. Today, a small remnant of this forest cover remains, largely hugging the most inaccessible mountain ranges, which make logging or conversion to plantation agriculture unfeasible. For foresters throughout the archipelago's colonial and postcolonial history, the primary culprit for forest decline has been not the voracious logging industry but swidden agriculture, known locally as *kaingin*. In addition to being cast as a threat to economically valuable forms of forest use, swidden agriculture in the Philippines has long tapped into a host of modernist anxieties surrounding backwardness, mobility, state authority, and fire use that have animated both biblical rhetoric of apocalyptic doom and large-scale projects of forest governance. Perhaps most problematically, the label of *kaingiñero* has come to encompass not only migrant-settler households attempting to escape poverty and landlessness by moving into still-forested areas but also the nation's indigenous peoples, whose lifeways, cosmologies, and socialities are often bound up with shifting agricultural practices. As a result, indigenous communities have come to be targeted by a series of forestry interventions that have aimed to extinguish *kaingin* from the uplands and, in many parts of the archipelago, are still popularly blamed as the primary drivers of deforestation.

Regional climatic data reveal that rainfall, the most salient meteorological feature of indigenous peoples' everyday lives and rain-fed agricultural practices, has been historically variable and irregular on Palawan Island and is subject to considerable year-to-year variation and uncertainty. Climatic data collected since the 1950s by the Philippine Atmospheric, Geophysical and Astronomical Services Administration (PAGASA) indicate erratic swings in annual rainfall that can vary from as little as 500 millimeters to 2,500 millimeters a year. Consistent with broader Philippine weather patterns, Palawan's intra-annual climatic variation is largely driven by monsoonal regimes that oscillate between the dry, northeastern monsoon (*amihan*) and the wet southwestern monsoon (*habagat*). Monsoonal strength and timing are significantly variable, in terms of both seasonal patterns and interdecadal variation in annual rainfall, as a result of shifting El Niño–Southern Oscillation regimes (Lyon et al., 2006). Due to the island's central mountainous spine, the east and west coasts of the island have distinct seasonal patterns, further complicated by the intense microclimates resulting from the irregular topography. In sum, Palawan Island's climate is extremely variable and unpredictable across spatial and temporal scales, and likely has been over the span of human occupation on the island (Bird et al., 2007).

By the time that anthropologists began working with indigenous communities on Palawan in the 1950s, the island's swidden cultivators had long culturally accommodated annual seasons of alternating extreme drought and flooding and year-to-year irregularity by incorporating climatic uncertainty and intensity into ritual practice and adopting livelihood activities that mitigate environmental risk (Fox, 1954). The "multifield, multicrop, multiphase" long-fallow swidden cycles of the island's indigenous peoples common before the 1950s complement unpredictability in rainfall and sustain food production amid oscillations in extreme weather (Warner, 1979, p. 166). Swidden-oriented ritual practices throughout the island also commonly focus on the control or mediation of uncertain weather patterns, and climatic conditions are widely the subject of special cultural elaboration and closely connected to cosmological balance in mythology and ceremony.

As elsewhere in the Philippines, swidden has increasingly been displaced from the center of daily social and ritual life, alongside the social and spatial marginalization of the island's indigenous peoples (Eder, 1987). What historical processes have resulted in this reorientation in livelihood practices and decline in relative economic and political power? Prior to the Second World War, indigenous population density was low, and households made their livelihoods by accessing resources from the coasts, rivers, and mountains throughout the island. Although regional migration from the Sulu Archipelago during the nineteenth century had a profound impact on indigenous peoples' social and economic practices in the south of the island, it was not until the 1950s that Palawan's reputation as a peaceful and abundant frontier attracted large-scale migration from more overpopulated, resource-depleted areas of the Philippines. As a result, Palawan's population grew progressively during the twentieth century, from approximately 6,200 in 1903 to over 400,323 in 1990 (Eder & Fernandez, 1996, pp. 10–11). These migrants quickly "filled in" Palawan's coastal plains, and by the 1980s the coastal forest that once dominated the lowlands of the island had been replaced by paddy rice, coconut plantations, and migrant fishing and agricultural communities (Eder, 1999, p. 28). Ethnographic accounts of frontier settlement have consistently highlighted how homesteading migrants and extractive industries (primarily logging and mining) leveraged control of newly established bureaucratic infrastructure, relative economic power, and coercion to rapidly claim and transform the island's coastal plains and, in doing so, alienated indigenous Batak, Tagbanua, and Pala'wan households from their ancestral territories in the lowlands. Left landless, many indigenous peoples retreated or were forced into the foothills and steep slopes of the island's central mountain range.[7]

These hilly and mountainous areas of the island—places deemed marginal by pioneering settlers and state surveyors because of their inability to support intensive agriculture—were claimed and labeled by the Philippine state as public forestlands (Brown, 1991).[8] This classification has meant indigenous livelihoods have been subject to successive waves of resource management by forestry departments, local government units, and nongovernmental organizations (NGOs) who have sought to reconfigure agricultural practices in the island's forest landscapes. Since the American colonial regime pacified much of the southern Philippines at the turn of the twentieth century, swidden practices on Palawan Island have been routinely pressured by forestry officials, who have prohibited the making of *kaingin* in state forests and sought to provision indigenous people with alternate means of living in forest landscapes to manage the island's wealth in timber, mineral, and forest products and, later, service conservationist agendas.

As early as 1914, "non-Christian *kaingiñeros*" in central Palawan were apprehended, fined, and imprisoned by American colonial foresters in order to preserve timber for commercial extraction (Sherfesee, 1915).[9] At the same time, ploughs and other tools of sedentary agriculture were distributed to indigenous people to promote residence and livelihood in the lowlands (Denison, 1915, p. 115). The 1970s saw a shift in Philippine forestry policy toward the limited recognition of upland dwellers' rights to residence and livelihood on state forest lands (Makil, 1984). However, in much of Palawan the period marked "the first effective enforcement of anti-*kaingin* laws" (Raintree, 1978, p. 31) as the bureaucratic infrastructure of the Bureau of Forestry (later the

Bureau of Forest Development) expanded into areas that previously had little or no formal enforcement of regulations prohibiting the clearing of primary or old-growth forests. These changes in enforcement put significant pressure on swidden practices, which were by now largely constrained to the uplands of the island.

Following the fall of the Marcos dictatorship in 1986 and the efforts of environmental civil society groups to secure greater rights to access forest resources for indigenous communities in the 1980s, Philippine forest management was progressively devolved through various legislative measures and tenurial mechanisms (e.g., Certificates of Stewardship and Certificates of Ancestral Domain Claim) designed to grant authority and responsibility to local users (Magno, 2001). While rhetorically oriented around community empowerment, in practice the devolution of forest governance has been seen to reinforce existing patterns of resource use and vested economic interests rather than improve outcomes for households that reside on state forestland (Gauld, 2000). On Palawan, community-based conservation schemes in the past three decades have primarily aimed to secure "alternative" (that is, nonswidden), market-based livelihoods for swidden farmers through the promotion of commercial agroforestry, irrigated lowland rice, or forms of handicraft production. While such programs aim to alter patterns of resource control and social difference, critical appraisals of conservation and development projects on the island suggest that these livelihood interventions and efforts to grant tenurial security to indigenous peoples on state lands have failed to alleviate, and in some cases have perpetuated, the marginalization of indigenous people (Dressler, 2009; McDermott, 2000).

Under this sustained pressure to remove or reconfigure agriculture production, indigenous swidden practices in the uplands of Palawan have declined in field size, fallow period, mobility, and crop diversity since the end of the Second World War (McDermott, 2000, pp. 356–66; Novellino, 2007; Warner, 1979). In many cases, these adjustments have been strategic responses to constraints in the ability to clear forests for cultivation in upland spaces. By sedentarizing production and minimizing clearing and burning practices, indigenous people may evade punitive regulations associated with forest clearing and, at least, ensure continued residency and partial subsistence security in their ancestral lands (Dressler & Pulhin, 2010). In other cases, new and less criminalized economic opportunities frequently draw household energy away from subsistence production—meaning households invest less in the making and maintaining of swiddens, forgoing time-consuming (and productivity-enhancing) activities such as burning and weeding (Novellino, 2007). The resulting downward pressure on agricultural productivity, seen in the relative decline of upland rice yields throughout the island, has further prompted many indigenous people to secure a large portion of their basic household needs through production for or participation in lowland markets.[10]

As swidden has been transformed or abandoned entirely within new livelihoods, many indigenous households are increasingly beholden to migrant traders for not only luxury consumption but also the goods now deemed necessary for everyday household reproduction—rice, kerosene, cigarettes, coffee, and sugar, among many other tradable products.[11] Commonly, accessing these necessities means reliance or specialization in a narrow range of commercial activities for cash income or goods on credit, or *utang*. This dynamic is most visible in the forest product trade, an important livelihood activity for indigenous households throughout the island, in which the squeeze on upland agricultural production has reinforced historic relations of debt and patronage and bound indigenous peoples in dependency to lowland economic intermediaries.[12] But the peripheral position of indigenous peoples is not limited to forest product commodity chains. Throughout Palawan Island, indigenous peoples have entered the live fish trade, handicraft production, or wage labor markets[13] on largely uneven terms with migrants who control the terms of exchange, exposing them to the negative impacts of resource decline and increasing debt burdens.

The classification of upland areas as state forestland, intensive forestry management efforts, and the overall decline of swidden within indigenous livelihoods are therefore central to understanding the marginality—and present economic vulnerability—of indigenous peoples within Palawan's contemporary social landscape. As political constraints on agricultural productivity have squeezed subsistence security, indigenous households have been incorporated into the bottom end of the island's extractive commodity chains. These trajectories of livelihood change and swidden decline now converge with the region's variable climate in Palawan's indigenous communities.

BARANGAY INOGBONG

Bataraza is the southernmost municipality on the mainland of Palawan Island and is sometimes positioned as a kind of "last frontier" of the "last frontier" of Palawan itself (map I.1). Its *poblacion*, or town center, was at the time of my fieldwork a dusty assemblage of mostly unpaved roads and creaky municipal buildings. Unlike areas of Palawan where "wondrous" natures are produced and enjoyed at massive scales by domestic and foreign tourists

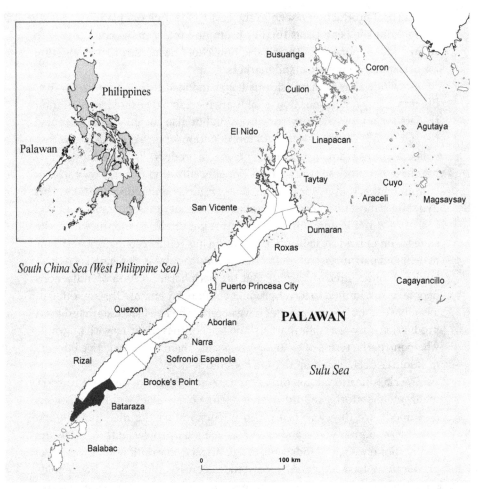

MAP I.1   Palawan Island.

(Webb, 2016), the municipality's rocky beaches and perennial low-level threats from Islamic separatists means that Bataraza is relatively unvisited. There are, on the whole, few attractions despite the efforts of municipal officials to cultivate interest in local environments and histories. Much of the Bataraza's limited prosperity is drawn from the nearby Rio Tuba copper-nickel mine, which contributes to the rough-and-tumble frontier atmosphere of the municipality and highly visible inequalities, where the thatched palm (nipa) huts of precarious wage laborers abut manicured golf courses and private helipads. This book's examination of climate and livelihood draws on the experiences of indigenous Pala'wan people residing in the state-owned uplands of the *barangay* of Inogbong. The *barangay*—the smallest level of administrative division in the Philippines—is only a five-minute motorcycle ride from the small town proper. The *barangay* of Inogbong is home to indigenous Pala'wan as well as Christian and Muslim households of migrant background, whose members trace their origins from throughout the Philippines. These families of migrant origin constitute the majority of the roughly four thousand individuals living in the *barangay*. The number of indigenous people there is difficult to pin down because of shifting residence and an unwillingness by census collectors to go into the uplands. I estimated roughly 100 indigenous households in the uplands during my research, though other sources place this at up to 160 (PTFPP, 2002a).

Much of southern Palawan stands in contrast to other areas of the island, where indigenous peoples and migrant settlers have moved more freely across social and spatial boundaries. The southeast coast of the island, and Bataraza in particular, is characterized by a sharp topographical and ethnic relief between indigenous, state-owned uplands and migrant, privately titled lowlands. The central axis of southern Palawan is dominated by the Mantalingahan Range, a series of high peaks that rise sharply from a small coastal plain in Bataraza and effectively seal it off from the Palawan Island's west coast across much of the south.

I had conducted research previously in central Palawan, where boundaries, identities, and livelihoods bled into one another in far more intimate ways (Dressler, 2009; McDermott, 2000).[14] In light of the work of other Palawan studies scholars who have cautioned against reifying the uplands as bounded and distinct spaces (Eder, 2006) or ascribing homogeneity to indigenous "communities" (Theriault, 2017), I was careful not to project a romantic isolation or purity into state forestlands. Despite this hesitancy, I was immediately struck by the visible and stark difference between the indigenous uplands and the flatter coastal plain when I first visited Inogbong and other *barangays* in Bataraza in 2010. These initial impressions were only reinforced over the course of my fieldwork in Bataraza from 2011 to 2012, as only three individuals living in upland hamlets self-identified as being *bisaya*. This term, though originating from the dominance of Filipinos from the Visayan Islands in early migrations to Palawan, is used by indigenous peoples throughout Palawan Island to refer to all Christian migrants, regardless of their origins. Bataraza is also somewhat unique in that it is poorly represented and serviced by Palawan Island's nongovernmental organizations concerned with environmental protection, many of whom cite the particularly violent and unwelcoming local politics. The only groups to venture into the uplands during my research were Filipino missionaries. Combined, these factors help lend a greater sense of separateness from the lowlands than the topography of Inogbong would suggest.

Though the boundary marking the end of privately titled land and the beginning of a space dominated by indigenous peoples is only seven kilometers from Bataraza town proper and the seat of municipal government, the uplands is marked by a conspicuous absence of the state. A rough gravel and dirt road runs up from the national highway to a small concrete dam that provides water for lowland households, but beyond the Inogbong River, which wends its way through the catchment and roughly bifurcates the *barangay*, indigenous hamlets are accessible only by foot or carabao (water buffalo, *Bubalus bubalis*). There is no electricity, and often-limited cell phone coverage due to the steep mountains that enclose the upper portion of the river valley. Across the river, the hillside presents a mosaic of a secondary forest growth, old and new swidden fields, small paddy fields, grasslands, scattered nipa huts, and large hamlets of Pala'wan households. After perhaps an hour or two's hike, at roughly five hundred meters above sea level, the landscape shifts. Forest cover becomes much denser, older, and steeper. Hamlets are much smaller and might consist of only two or three households. Livelihoods change, too. Swiddens are larger. Broken shrubs along the trail point to claimed wild bee hives. At eight hundred meters, a four- or five-hour hike from the lowlands and the limit of human residence in the area, the air becomes uncomfortably cold and shrouded in cloud.

Because of this steep terrain and lack of infrastructure, lowland Filipinos consider travel in the uplands—like indigenous life more generally—uncomfortable and undesirable. Despite the extremely nonviolent attitudes of most Pala'wan people and indigenous conflict resolution practices (with, as I explore in this book, some striking exceptions), many lowland Filipinos strongly fear traveling into indigenous areas. A Filipino salesman, traveling to Bataraza to sell wholesale pots and other cooking utensils at the local market, quizzed me about my research over my morning coffee during a stay in a lowland lodging house. He was a confident businessman, but he expressed genuine anxiety at the prospect of visiting an indigenous community for fear of being "poisoned"—a widespread sentiment among Hispanicized Filipinos on Palawan. Municipal civil servants and police also largely avoid entering into indigenous communities. One morning while we ate breakfast in the town market, my research assistant shared widely circulating rumors of a deformed indigenous baby born in the uplands of a neighboring *barangay*. The grotesque nature of the supposed disfigurement, a baby with the abdomen and legs of a pig, was enough to warrant a municipal police officer hiking into the community to investigate the claims. During all of my time spent in the area, this was the only account of state agents traveling into upland areas anywhere in the municipality.

This sociospatial cleavage in people, environment, and livelihood is a reality that many indigenous people navigate each day as they move from the forest interior to the migrant-dominated plains for wage labor in paddy fields (*basakan*), to sell cash crops and forest products and to purchase lowland goods at *sari-sari* (small, mixed goods) stores and the municipal market. In contrast to the fluidity that characterizes identity politics in other areas of the island, the ethnic boundaries between indigenous peoples[15] and Filipinos of migrant settler origin are relatively less malleable. This book is based on my ethnographic experiences traversing these ethnic and ecological gradients in Inogbong. This compressed variation meant that a day's fieldwork might include a dizzying array of activities: wending back and forth across the Inogbong River, inching across steep and slippery forest trails, spending time in my research assistant's church on Sunday mornings, visiting the *barangay* hall, and conducting household surveys in shrouded cloud forest.

In moving through these spaces to conduct interviews and observe daily life for the roughly one hundred Pala'wan households that reside on state lands, my focus was oriented around matrilocally defined hamlets ranging in size from approximately five to forty households. These clusters of kin can form the basis for significant economic, if not explicitly political, cooperation. The organization of schedules for communal labor, play, and ritual practice surrounding swidden agriculture take place mostly on the scale of the hamlet (figure I.2). Depending on size, hamlets will also generally contain one or two *panglimas* (customary leaders), who, anthropologist Charles Macdonald (2013) argues, have historically acted more as arbitrators or ritual specialists than as figures of coercive political authority or power. These hamlets are not necessarily fixed and enduring entities and, within local oral histories, are characterized by fluidity as village clusters fuse and fission according to shifting political and interpersonal dynamics. Pala'wan hamlets are, in practice, only loosely framed in terms of kinship, and the composition of villages is rooted more in individual residential preferences and livelihood aspirations that may or may not entail economic, social, and ritual cooperation, rather than tightly bound political units, as sometimes envisioned by forest management programs on Palawan (Macdonald, 2011).

Pala'wan livelihoods in Inogbong are diverse, and *uma* (swidden agriculture) remains a prominent though quickly changing livelihood for the vast

FIG. I.2
Pala'wan men operating a Malay forge

majority of indigenous households. Pala'wan people valorize rice as both spiritually potent and flavorful, but in reality, they rely on hardy and blander root crops such as cassava and sweet potato for basic subsistence. Cash income is increasingly important to sustain household needs and aspirations. A few ambitious households have sculpted irrigated paddy fields in what little flat or gently sloping land is available. Others rely on the sale of forest products, both legal and illegal. More commonly, indigenous people engage in poorly paid waged labor in migrant paddy fields in order to secure what are now considered basic household necessities: school fees, coffee, monosodium glutamate–based flavor sachets, clothes, a steady supply of rice, and medicines, among other goods. Social differentiation, though perhaps considered slight by the standards of lowland families, is palpably felt by indigenous households and marked by the use of galvanized iron sheeting for housing rather than palm or grass thatching, the use of generators, and confidence in dealing with migrant economic brokers.

These specific configurations of community, economy, and space in Inogbong are neither politically neutral nor static. Drawing on dozens of lengthy interviews and livelihood surveys from the state-owned uplands of the *barangay*, this book explores how they are shaped by multiscaled struggles over land and livelihoods that intermingle, in unexpected ways, with experiences of weather. In addition to drawing on ethnographic fieldwork with indigenous people in Inogbong, this book has been informed by oral histories from migrant lowlanders and interviews with *barangay* and municipal officials responsible for enforcing environmental regulations in forestlands. Because I lived and worked across the upland-lowland divide, often moving between lodging in Bataraza town proper and stays with indigenous families, participant observation has also informed my analysis of place making as an embodied practice of movement and often-daily feature of indigenous people's lives.

Of course, this book does not attempt to build an objective narrative of indigenous lives, livelihoods, and histories. It is a partial account based on complex intersections of gender and race. My positionality, as a white male "foreigner" in the Philippines, was relationally constituted with my primary research assistant, an evangelical pastor and indigenous man from a neighboring "tribal" group (the Tagbanua people). The Pastor accompanied me during much of my time in the uplands of Inogbong and for other research activities in Bataraza. His support was vital for my fieldwork in a range of ways: While most Pala'wan people are familiar with the Philippine national language of Tagalog, allowing me to casually converse and establish rapport with several families, his long-standing experience as an employee of a conservation NGO targeting indigenous people in other areas of Bataraza was invaluable in facilitating discussions in more distant areas of the *barangay*, where Tagalog is less readily understood. His affable nature put many indigenous *panglimas*, who came to respect his position and relative influence within the municipal government, at ease with my presence during my long and tedious interviews concerning the minute details of agricultural practices. Being accompanied by an indigenous man both opened new possibilities and closed down others. In most cases, my interlocutors were delighted to compare their ideas, practices, and histories to my research assistant's Tagbanua heritage and carefully review similarities in Pala'wan and Tagbanua language. Our own discussions of what he understood of Pala'wan ontology in relation to his evangelical beliefs, if not exactly harmonious, were a productive point for drawing out fine details of indigenous worldviews. His quick action after my ill-considered and disastrous attempts to ford the Inogbong River during one of the rainy season's many flash floods perhaps saved my life—though not several weeks of painstakingly collected field notes, consent documents, and much-needed eyeglasses.

In other cases, however, the gendered nature of our research partnership prevented us from gathering many women's perspectives of social and environmental transformation. While I worked mostly with a small set of families that became familiar and comfortable with our presence, sharing meals and long conversations, our journeys into further-flung hamlets were often marked by a greater tension. While indigenous men there were generally not shy in sharing their grievances or opinions of environmental change, women who rarely traveled into the lowlands and could not speak Tagalog well (if at all) were far less willing to share their opinions. The voices of Pala'wan women are therefore not absent from this study, but they are mostly restricted to the families I came to know well or to Pala'wan women in the foothill borderlands, the liminal zones between indigenous and migrant spaces, who are far more confident and outspoken. This is not to suggest that women in upland Inogbong do not have strong opinions and knowledge of agricultural production, governmental interventions, or climate change, only that these were often inaccessible to me. This book should be taken not as a comprehensive survey of Pala'wan life in Inogbong but a highly specific and partial account.

Social and environmental histories inform how the past is remembered or imagined and how it forms part of contemporary social life. Few of the Pala'wan people I interviewed, for example, were alive during the 1950s, when ritualized executions of incest offenders were said to still take place, yet informants in their thirties and forties enthusiastically evoke this period when discussing contemporary weather patterns. The most recent forest management project ended in 2002, but in 2011–12 this project's aftereffects were still manifest in local gossip over growing inequality. The state's power is experienced mostly at a distance, through rumors and brokerage. The 1997–98 El Niño, a distant event by many measures, is still discussed in reference to agricultural practice. "The past" is not simply a framing for academic analysis of contemporary practices, but a vivid element of life and livelihood for indigenous people themselves in Inogbong.

CHAPTER ONE

Making *Uma*, Imagining *Kaingin*

Conducting research in state forests in the Philippines often begins with a visit to the provincial or municipal office of the Department of Environment and Natural Resources—the state government agency that holds statutory responsibility for managing the "public domain" upon which many indigenous peoples in the Philippines reside. In early 2011, I made my own journey to the offices of the provincial Department of Environment and Natural Resources (better known as DENR) in the outer suburbs of Palawan Island's capital, Puerto Princesa, paying a "courtesy call" to explain my research to the organization whose turf I would be treading on. After presenting an introduction letter to the front desk and accompanying the office boy to the copy machine to dutifully collect my receiving copy, I was silently ushered through a series of waiting areas while they worked out just what to do with me. After I waited in a cavernous room of unused desks stacked high with paperwork, empty save for a solitary DENR employee slowly tapping out some unknown reportage on a typewriter, my handler pawned me off to an official. As I began to explain my project, it became clear that the DENR had little interest in my vague research ambitions in a far-flung *barangay* of the island. The official herself, justifiably wearied at having to listen to my elementary questions about forest policy, was far more concerned with adding a few small-denomination Australian coins to her international coin collection. As I stood to leave, she joked that they "have a real problem with the IPs—get it?" slurring the initialism for the officially sanctioned term "Indigenous Peoples" to make it sound like the Tagalog term for cockroach, *ipis*. With this characteristically Filipino wordplay, she eloquently flagged her exasperation with Palawan's recalcitrant indigenous swidden-practicing peoples, who, despite decades of efforts, refuse to act as they should.

These attitudes toward indigenous peoples—positioning them as stubborn and less than human pests—are by no means isolated, recent, or even hidden from public discourse. In 2015 and 2016, large El Niño–driven fires and drought ravaged some of the Philippines' most significant forests on the islands of Mindanao and Palawan. The visual imagery of scarce and valuable Philippine forests engulfed by fire provoked a substantial collective response to what was seen as a major environmental catastrophe. In media reports and policy statements over the course of the drought, many officials in the Department of Environment and Natural Resources and much of the Philippine national media placed blame on *kaingiñeros*—the marginal shifting cultivators who occupy the archipelago's uplands. As elsewhere in Southeast Asia, their use of fire was viewed not only as a direct source of forest decline but also as a dangerous ignition source for wildfires (Fowler, 2012). In the comment section of an article appearing in the *Philippine Daily Inquirer* in early 2015 that featured an image of mountainous landscape covered in patchlike scars suggested to be the product of swidden agriculture, one comment made by a Filipino netizen stood out: "The solution is simple: shoot to kill those farmers who are burning their farms. Eat their families to stop the tradition of *kaingin*" (Simple lang solusyon; shoot to kill mga magsasaka na nagsusunog ng bukid. Ubusin pamilya nila para itigil ang tradisyon ng kaingin) (Akramgolteb, 2015).[1]

While this is only one particularly emotive and anonymous comment, its extremity indicates the reductive nature of discussions of forest decline, which persists not only in policy and media coverage but also in everyday discourses surrounding environmental change and swidden cultivation. This quote, and my experience at the DENR, suggests a continuity between official rhetoric from the Spanish colonial period and the present and highlights the persistent role swidden and indigenous peoples occupy as "environmental destroyers" in contemporary anxieties over forest decline. Why does *kaingin* retain such a central focus in blame for environmental destruction in the Philippines? Why is it the subject of intensely emotive discourse that charges indigenous swidden cultivators with being not only environmentally destructive but also morally bankrupt and, frequently, less than human? What does their central place in narratives of blame "do" in Philippine environmental politics? In this context, different understandings and valuations of subsistence production and environment are reworked and embodied throughout the uplands of Palawan Island in contestations over the contours of indigenous livelihoods.

State forestry since the colonial period has framed *kaingin* and indigenous livelihoods in the Philippines as wasteful and destructive. These ideas regarding people and agriculture, developed in the context of colonial political economies in the Philippines, have had an enduring influence on the nation's environmental policies. Since the Spanish colonial period, the work of scientists, foresters, and policy makers has contributed to an official discourse surrounding forest decline that focuses on the role of indigenous peoples and their agricultural practices, thereby frequently eliding the role of commercial logging and other efforts to commodify upland spaces.

These ideas have shaped the specific policies that structure access to resources, governmental interventions, and indigenous livelihoods on state-owned lands. Through the production of policies focused on indigenous peoples, these persistent representations of indigenous peoples as forest destroyers have come to shape the postcolonial geography of the Philippine uplands. Social forestry efforts, heralded as not only a more humane break from previously punitive approaches but also a significantly more effective means of preventing environmental decline, obliquely perpetuated these representations and aimed to indirectly eliminate swidden agriculture by folding indigenous peoples into social and economic modernity. The rise of social forestry in the Philippines shifted the locus of governmental attention from the highly targeted and punitive disciplining of specific activities (i.e., *kaingin*) to the totality of indigenous peoples' social and biological lives, encompassing intimate thoughts and bodily practices. Forest policy rendered this expansive focus as the need to reform indigenous "livelihoods" while remaining indirectly oriented around expunging swidden from forest landscapes. This kind of thinking has profound implications for how community-based forms of forest management are designed and delivered, and it ultimately provides the context through which indigenous peoples residing on state-owned forestlands in Inogbong must make their livelihoods.

These persistently sedimented images of *kaingin* contrast with the role of swidden agriculture in Pala'wan cosmology, ritual practice, and social relations in Barangay Inogbong. The simplified representations of *kaingin* within forest policy contrast with the complexity of Pala'wan swidden practices, or *uma*. The successive stages that constitute the annual Pala'wan swidden cycle reveal how the community values and understands swidden agriculture as a practice concerned with more than the provision of food. Swidden is intimately connected to broader cosmological processes, smooth social relations, and local notions of tradition and heritage. The social and moral properties afforded to swidden agriculture are significant. This embeddedness means the Philippine state's efforts to remove or reconfigure the practice are best understood not only in terms of subsistence or productive capability but also as strongly moral and often explicitly "cultural" confrontations.

IMAGINING *KAINGIN*: SWIDDEN AGRICULTURE IN PHILIPPINE FOREST POLICY

The Philippines has a relatively unique colonial history in Southeast Asia. In a typical overview, the colonial history of the archipelago begins either with the arrival (and death) of the Portuguese explorer Ferdinand Magellan in the 1521 or with the establishment of the first permanent Spanish settlement on the island of Cebu by Miguel López de Legazpi in 1565. Following these events, the Philippines would endure several hundred years of half-hearted, and in many areas fairly tenuous, control by the Spanish, whose colonizing mission was led largely by the Catholic Church. Filipino historian Vicente Rafael (1993, pp. 18–19) has noted that the Philippines' position at the edge of the Spanish Empire meant that "for the vast majority of the natives throughout most of the three centuries of Spanish rule, the Spanish cleric ... came to represent their most tangible link to the Spanish *imperio*." Following the Spanish-American War of 1898, the archipelago was transferred to the control of the United States, which quickly, and ruthlessly, put down a nationalist insurrection that began under Spanish rule in 1886. After a brief period of military rule, a civilian government known as the Philippine Commission was established. In comparison to the Spanish regime, American colonialism was explicitly oriented toward promoting eventual self-rule, and therefore paradoxically sustained by efforts to produce the conditions for its own dissolution (Schirmer, 1975). While American rhetoric of the time was outwardly benign, the project of "uplifting" Filipinos entailed a rigorous and intensive project of reshaping the nation—while also occluding the continued and increasingly industrialized extraction of natural resources. American teachers who arrived en masse to fulfill this vision became, as colonial jurist in the Philippine Commission Charles Burke Elliot (1917, p. 229) noted, the "second army of occupation." As a result of these efforts, by the time that the US Senate endorsed the creation of the semi-independent Philippine Commonwealth in 1935 there were well-trained native bureaucrats, with high levels of English proficiency and comprehensive education, who could readily take control of state government.

Much of this narrative is a reductive story of Hispanicization, Christianization, and later economic development. In such explanations, diverse ethnic groups who fell outside of these processes, largely because of the tyranny of mountainous geography, came to be known as "tribal peoples," "natives," cultural minorities, and now in official discourse, "Indigenous Peoples." In many ways, however, the story of the indigeneity in the archipelago dovetails with the story of colonial forestry. The establishment of forestry institutions is bound up with the shape and nature of indigenous territories, the shifting political economy of making a living on "public forests," and the representational politics that have forged a close association between indigenous peoples and environmental destruction that continues to permeate NGO practices, governmental policy, and the national consciousness.[2]

Though Spanish colonial dominion of the Philippine archipelago developed over the course of the 1500s, "scientific" management of forests as a distinct category of land use began with the establishment of a formal forestry department, the Inspeccion General de Montes, in the middle of the nineteenth century. However, even prior to the establishment of institutions that produced knowledge about people and forests, swidden agriculture had come to occupy a focal point for Spanish efforts to transform colonial subjects. The work of anthropologist Felix Keesing (1962) in his detailed monograph *The Ethnohistory of Northern Luzon* provides important insights into how the emerging colonial state understood indigenous agricultural practices that, as early as the seventeenth century, were primarily associated with an emerging category of non-Christian peoples—variably termed *indigenas, tribus infieles, tribus salvajes*, and *igorottes*—and became bound up with a pejorative, moralized, and emotive official discourse. Keesing quotes, for example, an account from Spanish military captain Alinzo Martin Quirante's punitive expedition into the "unpacified" Cordilleras of northern Luzon in 1624 that labors on the unproductive and backward nature of shifting agriculture: "They attend to their waste and wretched fields in order to sow them with yams and camotes.... They do not have to plow or dig or perform any other cultivation than that of clearing the land where they plant" (p. 65). In contrast, Christianity, modernity, and progress became closely associated during this period with sedentary agricultural practices and tillage of the soil. Often facilitated by coercive military campaigns of relocation, the baptism and conversion of tribal peoples by the Spanish clergy took place alongside the reeducation of farmers to "plow and sow" rather than slash and burn (p. 74).

When the Inspeccion General de Montes was established in 1863, *kaingin* (also known as *cainges* in Spanish documentation) was quickly identified as a threat to the commercially valuable timber resources and ecological functions associated with tropical forests. The reports and published works of forester Ramon Jordana y Morera and forester and botanist Sebastian Vidal y Soler from the 1870s to the 1890s reveal how environmental narratives of forest decline and swidden blame were established and reproduced within the colonial imperative to commercially exploit the archipelago's resources. In these documents, the use of fire in agriculture was particularly associated with indigenous peoples in the rhetoric of the forestry service, which explicitly identified the difficulties of preventing non-Christianized people in the uplands from transforming forests into economically depleted grasslands. These narratives linked deforestation to varied forms of cultural difference. The expansion of troublesome cogon grasslands, or *cogonales*, composed largely of *Imperata cylindrica*, a rhizomatous and highly flammable grass species common throughout Southeast Asia, was blamed primarily on the swidden, ranching, and other fire-based activities of various non-Christian peoples. For example, in Vidal y Soler's (1874, p. 22) discussion of forestry issues on the frontier island of Mindanao within his monograph *Memoria sobre el ramo de montes en las isles Filipinas*, he notes, "In all the islands of the South, considerable quantities of good timber are lost that would be used for construction. The Moorish [Muslim] ranches move frequently, and they burn the forest in the place they have chosen to grow their crops for a couple of years. And the aboriginal Indians, always fleeing from their contact, are like their forerunners taking fire into the interior. This matter demands a preferential interest on the part of the local authorities, whose zeal should excite their superiors."

Unsurprisingly, despite nominal provisions to protect indigenous forms of resource use under Spanish colonial law, Jordana y Morera (1876, p. 11) notes that *kaingin* was formally prohibited as early as 1873, and the prohibition was later reinforced in the promulgation of the Definitive Forest Laws and Regulations in 1883 (Nano, 1939). However, despite the proposed "zeal" with which people and forests should ideally be managed, the vast size of the archipelago's forests relative to the small staff of the colonial forestry service meant that the impact of antiswidden measures was likely limited. The primary outcome was the entrenchment of intensely moralized environmental discourses that blamed the swidden of "tribal" peoples, rather than burgeoning commercial exploitation, for forest decline.

When American civil and military administrators turned their attention to the forest resources of their newly acquired colony in 1898, they found a detailed system of laws and regulations alongside a modern bureaucratic apparatus that, while drawing on scientific knowledge and forestry practices of European nations,[3] was crippled by a lack of resources and the inability of the Spanish military to exert effective control over much of the colony—particularly the southern portions of the archipelago, including much of Palawan Island. As such, it was not until the transfer to American civil governance in 1901 that the lens of professionalized scientific forestry, concurrent with American military power, was brought to bear on the tropical landscapes of the archipelago. When surveying the existing biophysical and regulatory landscape, American foresters chastised the lax Spanish enforcement that had led to an unregulated expansion of logging that was now seen to be deforesting the archipelago (Bankoff, 2009). Despite some attention given to poor logging practices, *kaingin* continued to provoke the moral outrage of the colonial forest bureaucracy. This frustration was vented in the production of technical and bureaucratic documentation of what had become the Bureau of Forestry. The documentation of the bureau, including annual reports, technical monographs, and materials in forestry journals, provides a rich trove of official material that connects indigenous agriculture to environmental change. *Kaingin* was viewed within these documents, largely without exception, as a destructive, wasteful, and often explicitly "evil" practice.

What justified this rhetoric that revealed swidden agriculture as the central anxiety of colonial foresters? Rather than being derived from the systematic study of indigenous swidden systems, this moralized discourse stemmed from early appraisals of Philippine forests, which were valued entirely in terms of the commercial potential of the archipelago's timber reserves, "minor" forest products,[4] and to a lesser extent, the potential for cash cropping in upland areas (Tucker, 2000). It was widely believed that scientific—or "practical"—management of forests according to American forestry principles could yield a sustainable source of income from the colony (Dressler, 2009, p. 53). In a statement indicative of the value that the Philippines held for its colonizers, Dean Worcester, a prominent member of the US colonial administration, suggested in 1914 that the forests of the archipelago were "money in the bank" (Bankoff, 2007, p. 231).

American foresters demonstrated specific anxiety surrounding the widespread cogon grasslands that supposedly covered, in varying estimates, between 20 and 40 percent of the nation's landmass. Colonial officials considered *Imperata cylindrica* economically inert. The grass was, in the words of Bureau Chief William Forsythe Sherfesee (1916), "useless for grazing . . . a fire menace, destroys the productivity of the soil in which it grows, and, worst of all, is the breeding place of the locusts which yearly swarm over the Islands." Drawing on widely circulating narratives embedded in the Philippine forestry during the Spanish colonial administration, American forestry officials saw human-induced fire as the primary means through which tropical forests were transformed into what they considered economically valueless grasslands. They assumed tropical forests to be naturally "immune" to wildfire (Ahern, 1902). In a clear explanation of the link between *kaingin* and forest decline that would come to have a lasting impact on forest policy, Harry Whitford (1911, p. 13), a prominent Yale-trained forester in the bureau's Division of Investigation, argued in a report produced for the Bureau of Forestry, "If the jungle growth is set on fire, as is frequently done, nearly all plants except the grasses are killed. In this way through many years vast areas of forest lands have been converted into cogonales, and repeated firings have prevented any change in their vegetation . . . grass lands are prevalent on land of nearly all types of topography, from sea level to the tops of mountains."

In the documentation of the bureau, foresters largely attributed blame for fires in cogon and forest landscapes to the livelihood activities of peoples now officially categorized as "non-Christian tribes." "Tribal" peoples set fires for a range of reasons, including the promotion of cogon regrowth for grazing,[5] as part of hunting practices, and simply for clearing land. Swidden agriculture, however, was identified as the most significant and widespread source of fire and the creator of vast cogon landscapes. The American botanist Elmer Merrill (1912, p. 150), who would later head the Philippine Bureau of Science, emphasized the role of swidden in producing and sustaining such landscapes, suggesting, "If the abandoned clearing becomes reforested, then it is only a question of time when the land will again be cleared for cultivation by the "caingin" system. . . . As the grass-covered areas become more extended, the fires often gradually push back the edges of the forest by the destruction of the young and the mature trees along the edges of the clearing, even when not aided by man." Merrill's analysis, like those of Whitford and other American scientists, was based not on detailed, long-term environmental histories of the archipelago but on observations derived from traveling, often observing forest landscapes only by boat—what the historian Greg Bankoff (2009, p. 371) has referred to as "forestry from the deck of a ship." These synchronic surveys of forest decline produced histories of Philippine forests that identified small-holder use of fire as the primary cause of deforestation and key to the maintenance of cogon landscapes. These observations ignored, for example, the rapid deforestation of Luzon and the Visayan Islands at the hands of Spanish-led logging, grazing, and plantations (Potter, 2003, pp. 39–40).

These assumed environmental histories meant that *kaingin* became identified as the central problem of forest governance in the Philippines, and in doing so, they elided the role of either Spanish exploitation or the rapid drive to modernize and derive revenue from forests under the American colonial government.[6] George Ahern (1910, p. 8), the first chief of the bureau for the American colony, noted in his annual report for the year ending in 1909 that "renewed efforts were made during the year to suppress the making of *caiñgins* in good forests . . . but it will be impossible to suppress this evil until a larger force is provided." In the following year, *kaingin* had become the "greatest problem that confronts the Bureau of Forestry" (Ahern, 1911, p. 8). Subsequent bureau heads William Forsythe Sherfesee, Arthur Fischer, and Florencio Tamesis echoed these sentiments in their annual reports. During the course of the American administration, the *kaingin* problem remained a central, yet unsolved, concern in forestry efforts, even toward the end of American rule. By 1930, the bureau's chief of the Division of Investigation continued to argue that "the evils of *kaingin* making are so evident and the problem is of such importance that its solution should be given immediate attention by tropical foresters" (Pflueger, 1930, p. 71). A key aspect in establishing the villainous nature of swidden was the circulation of images of scarred and depleted forest landscapes that could accompany the rich descriptions of devastation provided by the foresters. Photographic documentation of the destructive nature of swidden agriculture was a recurring visual theme of technical and scientific reporting of the time. These practices prefigured contemporary media reporting that graphically lingers on the impact of swidden fire on forest landscapes.[7]

Over the course of the US administration, a new cohort of Filipino foresters was trained in technical schools and colleges established by the colonial government (Roth, 1983, p. 43).[8] The American colonial government established a forestry school in 1910 and regularly sent Filipino foresters to the United States for specialized graduate education. Pejorative views of *kaingin* therefore became entrenched as the forestry service was progressively Filipinized. As the historian Richard Tucker (2000, p. 380) has noted, colonial forestry education "precluded a full understanding of the forest peoples and their lifeways" among Filipino foresters. As a result, the official rhetoric of new, Filipino foresters who came to prominence within the bureau in 1930s continued to frame *kaingin* as a "problem to be solved," and their narratives of forest decline emphasized the role of encroaching peasant farmers and indigenous peoples.

These Filipino foresters cited and reproduced the work of Whitford and other colonial scientists in ways that sustained the florid descriptions and pejorative view of swidden agriculture as a fundamentally destructive practice. For example, Filipino forester Juan Daproza (1931, p. 3), who earned a Master of Forestry degree at the University of Minnesota in 1925, wrote in an editorial for the bureau's staff publication, the *Makiling Echo*, in 1931 that equated *kaingiñeros* with insects and pests: "Prudent judgement demands that we inject the virus of forestry knowledge into the peoples [sic] blood so that they may become forestry minded and wake up to the moral and sacred obligation to reforest all idle lands, and perpetuate existing forests by wise use and close utilization and by effective protection from *kaingin*-makers, insects and other pests." These kinds of views were by no means unique. In similar bureau publications, Filipino foresters positioned *kaingin* as "one of the main cases of forestry destruction in the Philippine Islands" (Anonymous, 1931, p. 2–3) and extensively labored on the "evil effects of *kaingin*-making" (Dacanay, 1937, p. 22).

The Philippines gained independence from the United States in 1946. However, this independence did not result in a reevaluation of colonial attitudes toward upland peoples and their resource use. *Kaingin* largely remained fixed in political and public consciousness as the nation's foremost environmental problem, and foresters continued to blame local peoples for rampant forest decline.[9] From 1946 onward, the independent national government saw continuing economic value in the commercial exploitation of forest resources in the name of revitalizing a nation devastated by war—and these imperatives in turn supported already-dominant views of swidden within the Bureau of Forestry as "primitive and wasteful" (Gülcur, 1965, p. 96). Rather than faulting colonial exploitation, Filipino foresters in the immediate postwar period continued to see *kaingin* as "undoubtedly, the most harmful agency causing the destruction of the forests" (Dacanay, 1949, p. 3). In the introduction to the first issue of the *Philippine Journal of Forestry* issued after the interruption of the war, Bureau Chief Florencio Tamesis (1947, p. 3) clearly identified *kaingin* as a core focus of the renewed forestry efforts: "The problem of forestry destruction by illegal *kaingin* is ever present as a major problem of forestry in this country. This pernicious habit of clearing timberland for temporary planting is responsible for the loss of valuable timber and loss of soil fertility particularly in places approaching marginal land. To counteract this, adequate protection is essential, requiring more patrol work and effective cooperation of the municipal as well as provincial administration."

National media and government policy framed the forests of the newly independent nation as a crucial resource that could provide a path to economic development and modernity following the devastation inflicted during the Japanese occupation. Because of these commercial imperatives surrounding forests, the postcolonial government and forestry department passed harsh laws that explicitly criminalized *kaingin*, and until the 1970s the nominal goal of swidden policy was the complete expulsion or relocation of all subsistence agriculturalists from state-owned land containing commercially valuable timber reserves. Legislation, such as the Kaingin Law (Act No. 274) of 1901 and the Revised Kaingin Law (Republic Act No. 3701) of 1963, stipulated heavy fines and lengthy jail sentences for the unlawful destruction of what were considered "public" forests. Ongoing concern regarding agriculture in the uplands culminated in the 1965 National Conference on the Kaingin Problem, which, among other recommendations, suggested the forced resettlement of upland farmers to lowland areas (Scott, 1979, p. 60).

Under the Marcos dictatorship during the 1970–80s, many of these policies of expulsion and relocation from forestlands were tempered or recognized as unenforceable in practice—especially in light of enduring political instability in rural areas over highly concentrated land ownership. In place of imprisonment and fines as a means of regulating resource use, forestry efforts became increasingly focused on reconfiguring subsistence *kaingin* into market-oriented, sedentarized upland agriculture to preserve existing forest cover. National projects such as the Forest Occupancy Management Program (1974), the Family Approach to Reforestation (1976), and most significantly, the Integrated Social Forestry Program (1982) sought to work with farmers within a new regime of ostensibly more inclusive and "social" forestry (Sajise, 1998, p. 227). Under these and other social forestry programs, the willing labor of forest occupants (who were formerly seen as "squatters") was understood to be critical in solving upland environmental, political, and economic problems through a new social compact in which financial support, technical aid, and tenurial recognition were contingent on sedentarization and forest protection.[10] While the reframing of environmental subjectivities was also a goal of American forestry—varied forms of forestry "propaganda" were often deployed alongside punitive measures in order emphasize the importance of timber resources for nation building to remote tribal peoples—the shift to social forestry heralded a far more expansive need to reconfigure upland subjects in order to achieve conservation objectives. Such aims were explicit within the Ministry of Natural Resources' rhetoric surrounding the Integrated Social Forestry Program: "The program aims to mobilize forest resources for economic and social progress of the nation through the involvement of kainingeros [*sic*] and other occupants of the forest lands who shall be made into effective agents of the state in food production and rehabilitation of forest lands" (Sorbara, 1998, p. 59).

Accompanying this shift in approach was a more indirect and technical discourse surrounding swidden agriculture: emphasis moved from the "problem of *kaingin*" to the "problem of upland development," in which links between environmental degradation and poverty in ecologically sensitive areas replaced *kaingin* as the overt focus of forestry efforts. As a result, expansively conceived upland "livelihoods" became the locus of governmental attention. However, though these new projects were framed on the ground in terms of economic empowerment in the uplands for marginal households rather than directly focusing on swidden cultivators, outcome indicators continued to measure success in terms of the ability to curtail the mobility of *kaingiñeros* and transform farmers into productive land managers (Fujisaka & Capistrano, 1986, p. 225). Productivity, in these cases, did not mean subsistence security but the ability of upland farmers to contribute to production for national and international markets and act as sedentary tree planters. Many social forestry projects therefore obliquely sustained views of swidden agriculture as an environmental problem, as the objectives of a program targeting indigenous "negritos" in Negros reveal: "[The project] believes that the farmers have to pay the price—one that is reasonable and within their reach. Under the social forestry concept, the farmers are converted into forest protectors instead of forest destroyers. Their tenure in the area is not illegal as long as they protect the remaining forest in the area and integrate tree crops into their farms. The result is a social agroforestry system" (Cadeliña, 1986, p. 110).

Drawing on regional work on shifting cultivation systems by the United Nations' Food and Agriculture Organization (FAO) and later the then-named International Council for Research in Agroforestry, social forestry programs from the 1970s to the present made agroforestry the primary technical means to achieve the transformation from farmer to "forest protector." However, this emerging reliance on agroforestry was hardly neutral in its prescriptions of tree farming. In many cases, the sedentarization of swidden agriculture was an explicit goal of agroforestry projects that aimed to "take the 'shift' out of shifting agriculture" (Scott, 1979, p. 61) by recommending the intercropping of commercial perennial species with annual food crops. Under such imaginings of upland livelihoods, households would stay rooted in place and provision for themselves and the market through planting crops that were deemed, at the time, to be of high export potential.

In many cases, the transformation of livelihoods was conceptually implicit in the design of new ways of farming in the sloping uplands. For Hispanicized Filipinos who had migrated into upland areas, these prescriptions closely aligned to their agricultural aspirations to ultimately sedentarize and intensify their productive capabilities, but the prospect of largely sedentary livelihoods would have been challenging for many indigenous swidden-practicing communities, whose subsistence (and cultural and social life) depended on rotational practices. For example, forms of hedgerow intercropping or alley cropping, favored agroforestry methods in the 1980s that were widely implemented as part of the then-popular Sloping Agricultural Land Technology (SALT) package, were incompatible with many forms of swidden cultivation and required long-term investments in a relatively small area. This focus on demobilizing upland livelihoods through reordering agricultural practices was supported by new tenurial mechanisms designed to develop a sense of ownership and investment in single parcels of land—a quality deemed lacking in upland communities. Though sitting alongside community-based arrangements,[11] the primary and most common tenurial tool of social forestry projects in the 1980s was the individual Certificates of Stewardship contract, which granted farmers legal use of a single, fixed plot for twenty-five years, contingent on the improvement of land through tree farming (Lynch, 1986, p. 284).

At the same time, though there was some recognition that the nation's indigenous communities in the uplands were entitled to land rights distinct from Hispanicized households that had migrated from the lowlands into newly deforested areas, these "primitive tribes" and their equally primitive agricultural practices were seen as inherently threatening to forests (Makil, 1984). While some foresters engaged with anthropological research on swidden in formulating policy, these efforts were less concerned with building greater understanding of diverse and complex agroecological systems than reifying distinctions between the practices of still-primitive tribal peoples and the upland agriculture of transplanted Hispanicized peasant farmers. In policy documents focused on the enduring *kaingin* problem, "true" rotational shifting cultivation remained practiced by only the "least civilized cultural minority groups," who were diagnostically "illiterate and non-Christian" (Population Center Foundation, 1980, pp. 16–24). The specific kinds of *kaingin* practiced by indigenous peoples were seen as the most destructive of all kinds of upland agriculture due to the perceived tendency of tribal peoples to move frequently and favor clearing old-growth forests. Research on *kaingin* conducted by the forestry service, largely through the bureau's Forest Research Institute, which was established in 1974, was aimed not at building a more nuanced understanding of diverse upland agriculture practices but at leveraging social research as a means to gain the cooperation of recalcitrant *kaingineros* in agroforestry interventions. In the Forest Research Institute's monthly newsletter, *Canopy*, an official in the "socio-economics division" clearly articulated the role of research in curbing *kaingin* and remedying the deficiencies of upland peoples: "Some of the *kaingineros* are continuously exhibiting a negative attitude towards tree planting, lacking the needed initiative; they just go about their usual way of life.... Any development effort towards the rehabilitation of Angat Watershed or any other forest watershed for that matter, will be futile without the full cooperation of the settlers, *kaingiñeros* or squatters within the watershed. Understanding them is the key to obtaining their cooperation" (Calanog, 1977).

Following the ousting of the Marcos regime in 1986, understandings of indigenous livelihoods and environmental governance efforts became increasingly coproduced and codelivered by civil society and state actors. Over the same period, governmental policy and popular opinion have increasingly recognized indigenous peoples' rights to land and forest resources. The then-named Bureau of Forest Development was reorganized under the Department of Environment and Natural Resources (DENR). During this period, forest policy began to emerge from a complex interrelationship between NGOs and state agencies, as individuals would move relatively fluidly between organizations, bringing new ideas across institutional boundaries (Bryant, 2008). During this period, two key developments came to institutionalize livelihood intervention as the key means to protect forests. Firstly, in 1995 "community-based forest management" (CBFM) was officially enshrined as "the national strategy to achieve sustainable forestry and social justice" by the then-president, Fidel Ramos, through Executive Order 263. This legislation formalized trends focused on the provision of alterative livelihoods. The legislation explicitly mandated that the now-named Forest Management Bureau within the Department of Environment and Natural Resources would deliver "community training and empowerment, enterprise development, agroforestry development, tree plantations, and other non-forest-based alternative livelihood systems" (Executive Order 263, 1995).

Secondly, sustained advocacy from both within the DENR and civil society during the 1980s and 1990s culminated in the Indigenous Peoples Rights Act (IPRA) of 1997. The act provided unambiguous support for indigenous peoples' right to own and manage the forested landscapes they have historically accessed through new tenurial mechanisms for what are now known as Certificates of Ancestral Domain/Land Title. Critically, this legislation acknowledged the emic values and "indigenous knowledge systems" that local peoples may apply to their forest environments and livelihood practices. Included in the IPRA was also the provisional rehabilitation of indigenous environmental practices through clauses that granted "sustainable traditional resource use rights" (Republic Act 8371, 1997), manifest in the often haphazard, partial inclusion of indigenous peoples in management processes and in the formulation of so-called Ancestral Domains Sustainable Development and Protection Plans.

However, despite this trajectory of increasing recognition for the rights and emic values of indigenous peoples and other upland residents and a nominal concern for issues of social justice in forestry, *kaingin* remains implicitly identified as a primary cause of deforestation through a language of "alternate livelihood" provisioning and "forest stabilization." Many governance endeavors overlook the persistent importance of swidden agriculture for many indigenous communities—and, importantly, these devaluations of swidden continue to inform the practical interventions into forest landscapes that seek to limit access to forest resources and redirect households into approved forms of livelihood. There is an enduring gap in the uplands between how indigenous livelihoods are imagined in the work of government and how indigenous peoples themselves value and put into practice swidden agriculture.

MAKING *UMA*: GOVERNING ENTANGLED LANDSCAPES

In 1957, anthropologist Harold Conklin produced a report for the Food and Agriculture Organization that described, in extraordinary detail, the swidden practices of Hanunóo people in the Mindoro uplands, with whom Conklin had spent several years completing his doctoral fieldwork. Conklin's descriptions, which dwarfed prior efforts to study shifting cultivation in their detail and specificity, embedded each stage of the annual swidden cycle within Hanunóo cosmology, ritual, and social life. In his schema of swidden-practicing peoples, this embeddedness identified the Hanunóo as "integral," rather than "partial," swidden cultivators, for whom agriculture was tightly interlinked with other dimensions of life. In Conklin's own words, integral swidden systems were a "more traditional, year-round, community-wide, largely self-contained and ritually sanctioned way of life" (p. 2). While these classifications may seem reductive and poorly representative of the complex realities that upland peoples in Southeast Asia face today, considering embeddedness and describing practice in detail are vital to understanding how struggles over place can be fractured and messy, how vulnerability to weather is produced and managed by indigenous households, and how climate change is comprehended through narratives of self-blame.

On Palawan Island more generally, indigenous agricultural practices are largely referred to as *kaingin* in the everyday discourse of lowland Filipinos, employees of NGOs, and state forestry officials,[12] and this term remains held in the Philippine national imagination as a practice that is environmentally destructive, primitive, and destined for extinction. Knowledge about indigenous practices is frequently produced in places far removed from where such practices occur. During a visit to one the small islands off the coast of Puerto Princesa, our group's tour guide pronounced with certainty and authority that *kaingin* was now completely eradicated throughout the province due to state restrictions. Despite his proclamation, and although the city government had briefly banned swidden cultivation in Puerto Princesa City in the mid-1990s, swidden remains central to the vast majority of indigenous households on Palawan.

A question that is rarely asked, even in academic research focusing on indigenous peoples, is how do Pala'wan households in Inogbong understand and value swidden agriculture, known locally as *uma*?[13] That it is valued is not in question, but there is little consideration of the everyday specificity, multiplicity, and (sometimes) ambivalence that surrounds the practice of shifting cultivation. Pala'wan articulations of swidden value are sometimes framed through a politicized moral discourse: the right to cultivate in the uplands, the right to subsistence, and the right to survive. For example, if you were to ask the female or male household head why they continue to cultivate in the uplands despite poor fertility, uncertain climatic conditions, and unending pests, the most common response would concern the means of food procurement: "If there is no *uma*, how could we eat?" However, swidden agriculture is understood, valued, and embedded within Pala'wan ontologies beyond subsistence rights. Many of my informants suggest that making an *uma* is *adat et keguranggurangan*, a custom of the ancestors that, ideally, should not be abandoned or forgotten. Why and how is *uma* positioned within the political discourse of many Pala'wan as a "cultural" as well as subsistence or economic concern? Understanding the strong, though transforming, social and moral character of agriculture in the contemporary livelihood discourse in Inogbong requires careful examination of how specific elements of the annual agricultural cycle connect to indigenous cosmology, ritual, and "good" social relations.[14]

Site Selection and Clearing

The first stage in Pala'wan swidden cycles is selection and clearing of a forest plot for cultivation of somewhere between one-half and one hectare in size.[15] In the catchment's forested interior, state restrictions on felling in various types of forest limit the range of new plots to an extent, and Pala'wan farmers are keen to emphasize they do not clear large trees.[16] Clearing in mature secondary or old-growth forests is certainly not unknown, but it is surrounded by an atmosphere of risk. Members of the municipality's environmental enforcement group, the Bataraza Bantay Bayan (Bataraza City Guard), report that they employ informers among upland residents to watch over their neighbors and report illegal activities such as small-scale logging or clearing forest in the upper portion of the *barangay* to the municipal authorities. Though I found the actual existence of the informers difficult to confirm, such statements by state officials have helped shape elaborate stories among Pala'wan households of state-owned helicopters and other aircraft that vigilantly survey the activities of indigenous peoples to prevent the clearing of old-growth or mature secondary regrowth forest. Certainly, no Pala'wan household I encountered ever confessed to clearing in anything larger than immature secondary growth, though my own observation of clearing in the forest interior puts such claims into question (figure 1.1). However, old-growth or even mature secondary forest cover is not necessarily considered an optimal site for clearing by many households, particularly those who have come to invest less in swidden production. For example, the felling and drying of enormous trees is often seen as a difficult activity and the work of only serious swidden cultivators. The skills and collective labor required to clear such fields are often regarded by many Pala'wan men as a

FIG. 1.1
Household cluster and adjacent rectangular swidden clearing in the *barangay*'s forested interior

practice of their fathers' generation. Nevertheless, Pala'wan recognize the role of preexisting forest cover in ensuring a good harvest and understand the correlation between fallow periods, soil fertility, and weed prevalence. Cropping is understood to progressively exhaust the *taba et lugta'*, or fat of the land. When clearing a plot, Pala'wan therefore seek a balance between labor investment in clearing, projected harvest, and weeding difficulty.

Decisions regarding clearing forests are also mediated by the potential presence of malevolent, invisible creatures who reside within or near large trees (Theriault, 2017). Such areas of forest are "forbidden" to clear for agricultural production. The influence of these considerations is visible in the patchwork of isolated forest cover that partially constitutes the landscape of Inogbong.[17] These forest patches are indeed owned or claimed by Pala'wan households, but they choose not to cultivate them because of the beings who reside within the trees and, if disturbed, may bring sickness and even death. Ritual testing of forested areas is therefore considered essential before clearing, as larger trees could be inhabited by *lenggam* or *taw't geba* (people of the forest) (figure 1.2). Determining the presence of these forest people is a varied practice, but at its most elementary level it involves a conversation

FIG. 1.2
A Pala'wan man imitating a malevolent forest spirit

with spirits inhabiting trees within a potential clearing. This may require walking through the site and talking to any spirits present. If a portentous dream is experienced that night, the area is deemed an inappropriate place to make a swidden.

Once the *uma*'s location has been established, the plot is first cleared of small shrubs and vegetation. If larger trees are present, they are felled with axes and then further divided into small pieces to ensure they will be evenly dried and result in a complete burn. The precise timing of clearing is contingent on the extent of vegetation present at the field, as an effective burn of the plot requires cleared vegetation to be thoroughly dried but not yet decomposing. According to one informant, if the swidden is cleared from old growth, the vegetation and trees must be cleared in December–January in order to completely dry the processed tree matter. The time required for drying likewise decreases with the stage of forest succession, with cogon grass or brushland[18] requiring only a few days to a week of intense sunlight before burning.

Burning

The arrival of the southwestern *habagat* monsoon provides perhaps the largest source of uncertainty during the swidden cycle. At this critical juncture, the farmer must judge the correct time to burn, ideally a week or two before the arrival of the monsoonal rains associated with shifting wind direction. By this time any cleared vegetation in the *uma* plot must be dry but not yet rotting. The coming of the southwest monsoon ideally occurs in May but is subject to extreme variation. Misreading the arrival of the monsoon is believed to have significant implications for the swidden cycle and ultimately rice yields; if the monsoon arrives late and the *uma* is burned too early, the nutrient-rich ash deposited through burning may be swept from the field before planting can be completed or may facilitate the growth of weeds that require removal before planting. If left too late, the heavy monsoonal rains result in an inability to burn the field. The unburned and moist vegetation is simply moved off the field and cropping proceeds on schedule. Failed or poor burns, however, lead to vastly greater number of weeds and diminished rice yields.

Assuming an ideal onset of the monsoonal rains in May, the male household head will wait for the appropriate conditions for burning. After a few days of intense sunshine and a moderate breeze, the field may be burned.[19] During the first burn of the dried and processed vegetation, the farmer will prepare his firebrand (figure 1.3) and begin to burn the edge of the field closest to the headwind in a circular fashion, allowing the fire to

FIG. 1.3 A Pala'wan man burning his *uma* field

collapse in on and extinguish itself. Any remaining unburned vegetation is piled and reburned in a targeted fashion. Such practices vary, however, in line with the diversity of upland agricultural practices across households, but parallel practices of other indigenous groups on the island. Semipermanent upland cultivation or recultivation of the previous year's *uma* may require only a small amount of underbrush clearing and patch burning.

Cropping and Maintaining the Uma

After the burning is complete, the uncertain onset of monsoonal rains provides a continuing source of anxiety regarding when to plant, and specifically when to plant rice. Ideally, rice is planted in the swidden field one to two weeks before the onset of the rain. After heavy rains, the field becomes increasingly muddy and planting becomes arduous. However, if rice is planted too long before the onset of the monsoon, lack of water may damage the seeds, and weeds will have proliferated before the rice is established.

The crop composition of swidden field varies considerably among households according to dietary preference, commercial intent, and future plans for the plot, among other factors. Rice constitutes the focal point of planting practices, occupying the majority of labor requirements and ritual attention, though it is always accompanied by a suite of legume, root, and cereal crops that provide the bulk of actual subsistence. From my discussions with a wide range of households, I learned that *umas* may contain between two and eleven varieties of rice from approximately seventy varieties named by Pala'wan of the southeast coast. A diversity of varieties does represent, to some extent, an explicit risk-mitigation strategy, because *something* must grow from so many varieties. However, households also select rice according to a range of qualitative as well as quantitative traits, such as expected yield. Many households idealize diversity in flavor, aroma, texture, color, and ritual purpose when selecting rice varieties. For example, plots will almost invariably contain both glutinous (*maragket*) and nonglutinous (*marungras*) cultivars; while *maragket* is not fermented into ritually important rice wine, which requires *marungras*, it is necessary for postharvest thanksgiving offerings. Planting a wide variety of rice is also seen as a service to future generations or something of an obligation to Pala'wan households from their ancestors. These varieties that have been entrusted to them by their ancestors cannot be cast aside and should be preserved for their descendants to cultivate and derive benefit from.

Many Pala'wan suggest that a diversity of small, individual rituals should precede or accompany the planting of rice. For example, prior to planting, farmers may awaken early and scatter rice grains in each corner of their field to ensure a good harvest. Alternatively, rice grains may be tied to a trunk in the center of the *uma* and the trunk topped with a stone. More elaborate rituals are surrounded by debates over how often they are still performed, if at all, and by whom. According to some informants, they involve the construction of a small "house," or miniaturized rice granary, in the center of the field to hold the seed crop before planting. Whatever the action, small prayers are spoken in the field to the Master of Rice, Empu't Paray, deemed responsible for the growth of the crop.

The actual planting of rice grains is enabled through widespread reciprocal and nonreciprocal labor-exchange arrangements. Hamlet leaders or close kin will arrange intrahamlet schedules to facilitate quick rice planting. Several taboos that surround the timing of planting are based on the indigenous lunar calendar, and they often stipulate the completion of rice planting within a single day. Interhamlet labor arrangements are often nonreciprocal and, according to my informants, now considerably rarer than in the past, when rice yields were larger. If a particularly large *uma* is made by a well-regarded household, laborers from other hamlets may be offered food in exchange for the day of planting. The mechanics of rice planting involve the men first dibbling the earth with a dibble stick, followed by women placing a collection of rice grains in the resulting hole.[20]

Purple yam (*ubi*) is planted after clearing, but before burning, near the tree trunks. Millet (*aturay*) is planted before the rice and, unlike other crops, segregated from the rice on the borders of the plot. The remaining crops are planted with little collective or ceremonial effort; within a few days of rice, the corn is planted, followed by cassava and bananas approximately a week later, then finally taro and sweet potato. Rice varieties are segregated in sections of the field, with the remaining roots, legumes, and bananas (though not the millet) intercropped across the field. Plants such as beans, tomatoes, eggplant, and tobacco are often sown along the outside of the field. The exact combination and schedule of crops planted is, like rice, subject to a wide variety of demands. This array of options does not necessarily represent the typical swidden field but the possibilities through which households compose an arrangement of crops to meet a range of aesthetic, dietary, ritual, and commercial ends.[21]

Maintaining the *uma* through weeding, guarding, and the potential use of insecticides represents a capital- and labor-intensive period of the swidden cycle. Weeding begins approximately three weeks after planting. By this time, if not before, various grasses will have emerged from the soil. The process of weeding may take up to two weeks in a typical *uma* if undertaken by only one member of the household, or perhaps as little as one week if both husband and wife work together. The typical day of weeding lasts from 6:00 a.m. to 11:30 a.m., and then again from 1:30 p.m. to 5:00 p.m. In the intervening periods, vigilant households will construct a temporary hut near their fields in order to guard their crops against rats and bird pests. Insect pests such as *tayangaw* (possibly *Leptocorisa oratoria*), black bug, rice stem borers, and rice worms are also a significant problem for fragile rice plants and one of the most common explanations for poor yields. Many households therefore desire expensive chemical pesticides, but due to prohibitive prices, they are rarely used on a regular basis. More common are practices typical throughout the Philippines among poorer subsistence farmers, which include burning a range of materials (small crabs, animal manure, chicken feathers, and various types of tree bark) amid the field to ward off pests.

Harvests and Postharvest Fallow Use

The harvesting of various crops within the plot is staggered across the year. Rice, again, takes primacy and is constituted by several harvests, an importance that is emphasized by the extensive terminology employed for the varieties of rice and stages in harvesting. Short-maturing varieties (*inuna'an*) are harvested approximately two and a half to three months after planting—this first harvest is called *mandigsum*. Three to three and a half months after planting, the bulk of the rice is harvested (*dumasun*). Like the planting of rice, harvesting is often a communal affair, and similar labor dynamics come into play, often involving the exchange of harvests for labor, though this is less common now that rice yields have become trivial within many swiddens. This exchange may enable a diversity of rice for consumption, despite a household planting only two or three varieties themselves. The work group will comb the field, individually snapping the panicle of the rice with a small hand blade. The rice is then threshed, winnowed, and dried near the field or home. Rice harvests are stored in large woven baskets or plastic sacks. Milling occurs throughout the year, depending on when rice is needed.

In many hamlets, after the primary harvest, thanks are ceremonially offered to the Master of Rice and ancestors for giving Pala'wan people rice and teaching them how to practice swidden. Bamboo tubes containing glutinous rice and coconut milk (*lutlut*) are heated over fire, and small ritual sacrifices of rice, corn, and cigarettes or tobacco are made along with prayers of thanks for a bountiful harvest (even when harvests have been poor). Each household in the hamlet contributes a certain portion of rice for the event (perhaps one-half to one *ganta* of rice, or approximately two and a half kilograms). Even if there has been effectively a total failure in rice harvest (a not-uncommon event, according to oral history), in some hamlets Pala'wan households suggest they feel compelled to purchase rice from the lowland markets to contribute their share.

Finally, rice fields may be gleaned from October to January of the next year. Because of the small quantities involved, gleaning is generally a solitary affair, in contrast to the communal labor involved in the primary rice harvests, and may be completed by a single member of the household. Crops other than rice are harvested in a similarly staggered manner, with root crops in particular being harvested as needed rather than in a single event.

The postharvest use of the *uma* often varies according to the length of previous cropping. Frequently, a plot cropped for only one or two years may be recropped without fallow for a second or third year. These plots will instead be cleared of weeds and crop remnants and often accompanied by subsequent patch burns around remaining perennial crops (e.g., banana). If the soil's fertility has been exhausted, the plot may be abandoned to fallow, but it remains an important "bank" of food. For example, a household will often have access to one or more older plots with remaining root and tree crops. Cassava, in particular, is often planted to excess and therefore rarely totally consumed before the start of the following agricultural cycle.

Cassava has a long "life," and fully grown tubers are more resilient in the face of extreme weather—though with age their taste and hence desirability are significantly diminished. Older *umas* are therefore tapped largely in times of subsistence crisis, alongside the harvest of wild fruits and tubers. In other cases, cash cropping of, primarily, bananas and *kalamansi* lime (*Citrofortunella microcarpa*) has become an important livelihood trajectory for some Pala'wan households in Inogbong. Old swiddens are therefore often planted with perennial cash crops and converted into plantations after the final rice harvests.

Uma *in Pala'wan Cosmology*

Anthropologists working with Pala'wan people throughout the island's south have pointed to a cosmology that bifurcates the world into a realm of concrete material relations, through which humans routinely navigate, and an invisible realm of malevolent and benevolent spirits (such as the Empu't Paray, or powerful ancestor spirits) who dwell within and beyond the visible landscape (Theriault, 2017).[22] Across the uplands of southern Palawan, Empu' Banar[23] (the True Master or God) sits at the head of the pantheon of various "masters" or deities of all natural phenomena—rice, ants, flowers, etc.—who reside in the heavens. Each realm impinges upon the other, most tangibly manifest in the control of natural phenomena by their associated spirit masters. While each master is responsible to a large extent for the ongoing functioning of their respective wards, the control of climate and the broader "metaecology" (encompassing events such as landslides, sea level rise, and tidal waves) of the world is seen to rest with Empu' Banar, who is framed as the ultimate recipient in many sacrificial practices.[24] This notion of a landscape inhabited and controlled by a range of spirit entities with significant power remains even among Pala'wan in Inogbong who identify as Christian. Perhaps because of the erratic missionary presence in the area, there is little dissonance in identification as Christian and either the approval or the practice of indigenous rituals and ceremonies.[25]

While Pala'wan people engage with this nonhuman world for a variety of reasons, agricultural practice is a core concern that warrants sustained and ongoing social relationships with the unseen world. Dreams in particular are a key means through which both benevolent and antagonistic spirits engage with humans. For example, in everyday dream events, malevolent spirits visit those who disturb their homes, deceased in-laws haunt and sicken their sons-in-law for not fulfilling their obligations, and ancestor spirits visit in the form of pests to proscribe the use of pesticide. These kinds of accounts were so frequent that rarely would a week pass in the field without the unprompted mention of personally significant dream events from informants. However, while dreams provide a means of communication from spirit entities, humans are unable to negotiate these encounters to their benefit. Instead, the sacrifice or offering (*ungsud* or *simaya*)[26] is the medium through which human influence is exerted from the visible landscape to the realm of invisible spirits, and in Inogbong, the largest and most regularized ritual sacrifice is made or discussed in relation to swidden agriculture or facilitating the ideal environmental conditions for *uma* production.

The position of *uma* as an important focal point within Pala'wan cosmology can be readily illustrated through the ritual practices that surround the production and consumption of rice in upland plots. How many Pala'wan households in the uplands ritually engage with rice as a spiritually potent substance and a valued source of subsistence provides an example of how making *uma* and related practices are enlivened with the intertwined dependence of humans and nonhuman entities. However, the ritual use of rice as a powerful link between visible and invisible realms vis-à-vis swidden agriculture cannot be understood divorced from its divine origins in Pala'wan mythology, which provides a logic for a range of ritual practices. The myth relating to the origin of rice was the most readily recounted story during my discussions with Pala'wan people and is often used as a rhetorical device by politically active men to illustrate the persistence of upland agriculture in the face of state criminalization and poor yields. Beyond political use, this origin myth highlights the swidden field itself as terrain that is central to Pala'wan ontologies. It identifies the Master of Rice as an important figure in Pala'wan mythology and highlights the role of rice as a potent substance in sustaining not only human bodies but also broader environmental conditions that facilitate further practice of swidden. In the version of the myth presented below, recounted to me by a *panglima* of a hamlet in the foothills of the *barangay*, the Empu't Paray visits the husband of a household in a dream and instructs him to kill his son, whose spirit and substance now inhabit and facilitate the growth of rice as a crop distinct from other sources of subsistence:

> I don't know what your grandfather taught you, but this is what our grandfather taught us—in the past there was no rice when doing *kaingin*.
>
> They told us about the legend of rice. Now, I don't know where the woman went, but she had one son, and she said that she would go and find some viand. She got her net and went to the river. Her husband was in their house. Then the husband

went to sleep. When he woke up, he felt that something had sprouted on him.

The sprouted plant said, "Bring your son to the *kaingin*. . . . You must kill him there."

Then they went to the *kaingin*, and the son asked, "What are we doing here, Father?"

His father said, "Nothing. I'll just get thorns of *bugtong* [thorny rattan]." They brought one dog, but that was a beautiful dog, very good at catching pigs.

After arriving there, the child was continuously beaten with thorns. Then he dragged him around the *kaingin*. Through all of it, they dragged that person. Then he killed the dog. He put the tail like this in the side [of the plot], around there. Then he cut the feet, even the head, and buried it.

The mother came home and said, "Where is our son?" She asked, "Where is he?"

The father said, "At the river."

She said, "Where is he now? It's getting late."

He said, "Come on. Let's eat."

The woman started to cry. She's still looking for her son. After seven days she said, "Let's go to our *kaingin*. What we will eat is like a banana but with more seeds."

But after at most seven days they saw something sprouting, that was the *aturay*. . . . The field was encircled with *aturay* [from the dog's tail]. Afterward *gabi* was seen sprouting [from the head of the dog]. There was *gabi* sprouting, and we call it the good-smelling one. There's already bananas growing, but when they examine it, they found the black leaf in the middle, and that was the blood. It's the black seed of the rice, the black leaf. You have probably seen it before. That's the sticky one.

[In total] those are the *kinina, gandinga*, the *tawaran* [rice varieties], there were three. The *gandinga* has no black. It is pure and the first to grow there. That was a person [the bones of the person], the owner of the rice. That's what they [our ancestors] told us.

Afterward, when the rice was ready for harvest, the father said, "This is rice, let's get it."

"What will we get?" said the mother. "Anger?"

Then when they're cutting the rice, it spoke and said, "Oh, Mother, don't cut me. It hurts!"

She dropped her tool.

Then the man won't let his wife do the harvesting. He and his men will do it. In their home, he said, "Go and get the *tuway* [shell, used for threshing]."

"Ah," it said. "Don't scratch me, Mother. It hurts."

The man took it from his wife and said, "I'll do it. It's nothing. Okay, just dry this."

Afterward, she heated the rice in a pan. "Ah," it said. "Don't put me in heat. It's hot."

She said, "What is this?"

The husband said, "Let me do it." They had dried plenty of it, and then the man said, "Now that it's dried, you will mill it."

Now when she's milling, it spoke like this: "Oh, don't you pound me. It hurts!"

So then the man took over the milling, until it was ready for boiling. It was never mentioned how much rice they cooked, but they had cooked rice.

They said, "Okay, let's eat and see what it tastes like." Upon getting the rice, they said, "Aromatic!" They ate.

"Ouch! No, Mother," it said. "Don't bite me! It hurts."

The man said, "Just eat."

"I don't want to. This talks," said the woman. When it stopped talking, then [the] mother ate. And that seven *legkaw* [rice granaries] can't contain all the rice.

They harvested much of the rice, then the father said, "We still have more left." The stem of the rice was not thrown away but eaten. The stem is placed in the pot, added with water, covered, and put on the fire. When water was absorbed, then the rice will rise.

And so our fathers told us, "Whose rice wouldn't get mad?" It's not only a plant but it also refers to a person. That's why they said, "If you don't get consent from the owner, you might only get the *binhi* [only get your seed returned but no surplus].

We cannot let go of this belief because it's hard for us to just forget the remembrance of our fathers. If we would let go of it, our rice may grow differently. If you just knew that rice is really a person, because I believe that it is. Consider the *tawaran* [variety of rice], before drugs came. Whatever the illness was, if *tawaran* was used as medicine, the illness will disappear.

Informants suggested that rice, while rarely used alone, is perhaps the most important substance in sacrificial practices in Inogbong. Because of its divine origins, it is the only element that facilitates communication between the seen and unseen realms.[27] Swidden rice is therefore valued not only as an ideal source of subsistence but also as a form of divine spiritual currency in enacting human agency beyond the visible landscape.[28] It is also an essential element of the ongoing cyclical reproduction of the agrocosmological order. Pala'wan descriptions of weather emphasize that the ritual sacrifice of rice is necessary to create the ideal environmental conditions to facilitate future *uma* production.[29]

The annual making and performing of thanksgiving *lutlut* is the most regularized sacrifice that reproduces the moral-environmental conditions that facilitate future agricultural production.[30] It is conducted each year after the primary rice harvest within each hamlet. The term *lutlut* refers to both the material focus and the process of the ritual; members of the hamlet place glutinous rice and coconut milk within bamboo internodes and cook them directly on smoldering charcoal. The subsequent consumption of the *lutlut* is accompanied by the symbolic presentation of rice and other goods and the giving of thanks to the Master of Rice for the rice being consumed and offering hope that the deity will ensure ideal conditions that will facilitate the next year's harvest. Thanks may also be given to ancestor spirits who left living Pala'wan the skills and technical capacity for making swiddens and planting rice.

The ritual sacrifice of rice also takes place in response to specific environmental conditions. For example, small, personalized ritual sacrifices are made to ancestors who either possess the power to transform specific aspects of the landscape or act as mediators to Empu' Banar. Such rituals take place at the graves of powerful deceased relatives, where rice, oil, and incense are offered in various configurations. While these kinds of rituals have potential beyond mediating environmental transformations, they are seen to be able to temper extreme climatic conditions. The choice of rice employed in such rituals is linked with the type of change that is required. For example, in requesting rain the rice will be black, a representation of blackened, water-laden clouds. And inversely, if sunshine is desired during unseasonable rain, the rice must be white. During the time of my research, this type of offering had apparently not been performed for several years, as there was some debate as to whether climatic conditions were truly unbalanced enough—the Pala'wan men who had the ritual knowledge and ancestral linkages to make such sacrifices were, to some, overly cautious.

Larger-scale sacrificial rituals necessitate larger contributions and transformations of rice. The production of rice wine requires, at least, an entire sack of rice and is employed when asking for assistance from powerful deities to remedy serious disturbances in natural phenomena (usually made in reference to the health or productivity of an *uma* or rice harvest). Husked and unhusked rice is cooked and mixed with yeast in a ceramic vessel to ferment for approximately two months. When fermented, the rice wine is drunk through bamboo straws and accompanied by the continuous playing of gongs by men and the dancing of women in order to lure the spirit beings to the site of ceremony. Once they have been lured, verbal requests can be made to the Master of Rice or ancestor spirits who mediate requests for aid to Empu' Banar.[31] Many Pala'wan emphasize that such rituals are socially and materially demanding to arrange, as they can potentially involve dozens of participants. Thus, although this kind of sacrifice is very much a living tradition, it has been performed only twice in the past two decades.[32] These ritual practices not only involve the web of human relations through which swidden is practiced (in, for example, organizing communal labor schedules) but also operate through a broader agrocosmological order. Humans are not the only entities imbued with demands and responsibilities in the upland landscape: successful swidden production involves the consultation and cooperation of a household not only with other households but also with a range of nonhuman actors.

CONCLUSION

The dominant environmental discourses that surround *kaingin* in the Philippines have positioned swidden agriculture as destructive. Ideas regarding agriculture, livelihood, and ecological processes in the archipelago's uplands have endured through broad transformations in Philippine forest policy over the twentieth and twenty-first centuries, albeit in ways that now somewhat more indirectly identify swidden as an environmental problem, in contrast to the direct moral outrage that characterized the language of foresters until the 1970s. Despite increasing recognition of the rights of indigenous peoples and other uplands residents and some tentative efforts to recognize indigenous peoples' environmental stewardship, forest management efforts to remove or reconfigure swidden agriculture continue to unfold in the state-owned forestlands of Palawan Island and beyond.

In contrast to the enduring state imaginings of *kaingin* as solely a destructive and backward practice, most Pala'wan continue to see swidden, the making of *uma*, as a valuable practice that speaks to both critical subsistence and the cultural politics of tradition and ancestral connection. Although swidden is emically understood as being in decline or transformation, making *uma* remains vital to ideal forms of sociality and ceremonial activity in Inogbong. It is the focal point of much of the communal labor and interhousehold interaction through planting and harvest schedules, the most frequent ritual and ceremonial activities, and human interaction with the supernatural realm of benevolent and malicious spirits. Although swidden agriculture does not dominate all domains of life entirely, symbolic properties associated with *uma* production loom large in everyday life in the Inogbong uplands and frame the ways in which changes in social and agricultural life are understood. This significance points to how the governance of livelihoods can have political and ontological implications beyond questions of subsistence.

The *kaingin* plot is a terrain through which different understandings and valuations of subsistence production and environment are contested, reworked, and embodied throughout the uplands of Palawan Island. The persistent gulf between authorized state environmental knowledge and local realities has been an enduring feature of Philippine forestry and exists as a contested epistemological space through which visions of people and place are imagined and come to shape the very matters of survival for communities who find themselves the targets of increasingly fine-grained efforts to reconfigure their lives.

CHAPTER TWO

Rooted in Place

IN 2009, the proclamation of the Mount Mantalingahan Protected Landscape (MMPL) enveloped much of southern Palawan Island's central mountainous spine in one of the Philippines' largest terrestrial protected areas. Spanning five municipalities and covering over 120,000 hectares, the MMPL was a high-profile project aimed at preserving what were seen as the last remaining tracts of contiguous forest cover on the once-verdant Palawan Island. This proclamation renewed scrutiny of indigenous peoples in Inogbong and sharpened focus on the work of researchers, approval for which would now be required from the MMPL's managing board. In early 2011, I presented my research proposal for consideration and approval at a scheduled meeting of the MMPL's Protected Area Management Board (PAMB), held at a newly constructed convention hall in the booming southern town of Brooke's Point, only thirty minutes away from Inogbong and in the shadow of the nearby Mount Mantalingahan range. The meeting was a lively and well-attended event, with expansive representation by the large PAMB composed of a bewildering array of special-interest representatives, such as the officers of the Armed Forces of the Philippines and Catholic nuns, with debate and planning that conjured images of a clearly defined protected area. The board largely humored me and my research on the politically neutral subject of indigenous *kultura at kasayayan* (culture and history). Still, the mayor of Bataraza, who was at the time part of the PAMB's executive committee, gently interrogated me after the presentation: "Will you be collecting samples as part of your research?" In seeking to diffuse the tension, I joked, "No sir, my research is *chika-chika lang* [only chit-chat or gossip]."

In the weeks following this meeting, as the Pastor and I began to travel regularly into the uplands of Inogbong to visit with Pala'wan households and broker permission for my "*chika-chika*" in the area to begin, I saw no official markers that identified the beginning of the protected area and the end of privately titled land. Having previously visited the Puerto Princesa Subterranean River National Park, a protected area and important tourist attraction located in the center of Palawan Island, I had anticipated a suite of technologies that could facilitate state territorial control at high-profile conservation areas: prominent maps that showed the relative size and shape of the protected area, well-defined boundaries to manage and secure flows of people, forest guards to patrol the borders of protected and nonprotected areas, and a clear understanding from residents of what belonged where. As I moved across elevations, none of this was evident in the area. How could conservation territories, such as the emergent MMPL, be produced in the absence of functional boundaries?

Though protected areas and other conservation spaces often remain conceived as "power containers," the notion of a territory as a "bounded space under the control of a group of people, with fixed boundaries" is a highly contingent vision of the relationship between power and place that has largely existed only in the imagination of Western political theory (Elden, 2013, p. 18). The absence of boundary-based enforcement, while partly accounted for by a perennial lack of capacity that plagues many protected areas in the Philippines, also points to a different model of spatial production. This model maps awkwardly onto ideal visions of territorial control, in which power emanates largely from the state and is dependent on the strict enforcement of borders and boundaries. Instead, the production of upland spaces has been the result of contested and diverse processes over time, beginning in the precolonial period and shifting into new forms under the recent rise of community-based interventions that now dominate forestry efforts in the Philippines and much of Southeast Asia. Instead of utilizing clearly defined boundaries that organize human conduct, the production of conservation spaces in southern Palawan since the 1980s has taken place through fractured and uneven attempts to discipline indigenous minds and bodies and, in doing so, endeavored to imagine and reform indigenous lives to root them "in place."

Philippine forestry since the 1970s has mirrored, and many Filipino foresters argue pioneered, a global transition away from punitive measures of forest management to nominally more inclusive forms of environmental governance. However, community-based approaches and the provision of "alternative livelihoods," formalized in law as the national approach to forestry in 1995, have become sites where older prejudices about indigenous peoples have been reproduced and continue to inform forest policy that is outwardly concerned with issues of equity and indigenous rights. The work of changing forest landscapes through transforming livelihoods is dependent on the discursive construction and practical remediation of perceived economic malaise, poor education, patterns of residency, structures of leadership, and community health in indigenous communities. Community-based programs are therefore animated by efforts to regulate indigenous populations in terms of their everyday livelihood activities, desires, aspirations, and education in order to obliquely produce distinct and ordered conservation territories.

This kind of indirect territoriality has implications for the production of conservation spaces and the lives of indigenous peoples. Project and policy documents, oral histories, and ethnographic work with Pala'wan people in Inogbong document the implementation and enduring impacts of an early community-based forest management project, the European Union–funded Palawan Tropical Forestry Protection Programme (PTFPP), which operated in the *barangay* from 1995 to 2002. One of the key aims of the PTFPP was to create a finely graduated conservation landscape in which human activity was progressively restricted, culminating in an untouchable "core zone." Instead of creating strictly enforced ecological borders, the project aimed to reorder human occupation and land use through interventions that encouraged indigenous people to adopt less mobile lives and livelihoods. Though the PTFPP ended in 2002, its ideas and material legacies remain a vital aspect of indigenous people's lives and continue to inform environmental discourse and decisions about how, and where, to live.

FRONTIER SETTLEMENT AND THE PRODUCTION OF DIFFERENCE (1700s–1980s)

Southern Palawan has a history of colonial frontier settlement that continues to shape the marginality of indigenous people well into the present. In the late eighteenth century, Tausug communities affiliated with the Sulu Sultanate, a Muslim polity that effectively resisted Spanish colonial domination across the southern Philippines, had established themselves on the coasts of southern Palawan and integrated the island's indigenous peoples in broader trade networks known as the "Sulu zone" (Warren, 1985). In doing so, Tausug leaders, or *datus*, increasingly excluded Pala'wan people from much of the coastal zone (Macdonald, 2007, pp. 12–15). Southern Palawan had sustained Tausug trading settlements and administrative centers that, while small compared to the sultanate's capital on Jolo, were vital in ensuring the flow of goods and people to and from Palawan Island. Hegemony over the flat, coastal plains not only facilitated the sultanate's external trade with the British Empire in forest products (destined ultimately for Chinese markets) but also allowed a yearlong supply of tribute rice from upland swiddens to flow to Jolo island (Warren, 1985, p. 95). While Tausug leaders forcibly elicited much of these goods as tribute, forests products and rice were also often exchanged for salt, metal tools, and ceramics—essential elements of upland social and economic life (pp. 137–38). However, while southern Palawan was one of several major trade centers for the sultanate and eventually the seat of the sultan himself during a period of internal political strife from 1886 to 1894 (Ocampo, 1996, p. 31), the forested hills and mountains were of interest to the Tausug only as a source of rice, commercial forest products, and labor to be extracted, rather than as a space to be transformed and governed. In southern Palawan, this flow of goods from households residing in the deep forest interior was not elicited by direct territorial control of upland spaces by Tausug chieftains but mediated through Pala'wan residing among or near coastal Muslim settlements.

Lowland Tausug *datus* maintained this coercive control of the coastal zones of southern Palawan from the eighteenth well into the twentieth century. Despite formal recognition of Spanish authority on Palawan in 1886 by the now-in-decline sultanate, the transfer of the Philippines to the control of the United States government following the Spanish-American War of 1898 did not translate to the end of Tausug authority. Annual reports of Dean Worcester, the secretary of the interior for the American Insular Government, indicate that the new colonial administration had yet to establish exclusive military authority over southern Palawan as late as 1909. Though faced with overwhelming American military advantage, Tausug *datus* still attempted to maintain control over the flow of upland resources by robbing and killing European and Chinese merchants. Agriculture was uncommon among Tausug communities on Palawan, who subsisted primarily on the tribute elicited from Pala'wan swiddens (Worcester, 1909, pp. 126–27). Oral histories in Inogbong indicate that as late as the 1950s, lowland Muslim elites retained coercive control over Pala'wan labor and upland resources. As one early Visayan migrant recounted, Datu Sapudin Narrazid (who would later become the first mayor of Bataraza) may have also coerced Pala'wan households to work in lowland copra plantations: "You know the natives would only get fed and no wages. For one day's work they eat only once, without coffee. The *datu* is like their God. If there's work, the *datu* just hits an enormous gong, and when their *panglima* hears it and comes down to ask if there is work to do and then brings them [the Pala'wan] down."

Because of these ongoing conflicts in the south of the island between the now-permanently settled Tausug *datus* and the American colonial administration, the migration of Hispanicized and Christian Filipinos to Bataraza remained relatively limited prior to the Second World War. While much of the island had been subject to sustained, though minor, migration since the turn of the century, it was not until the 1950s that early Christian settlers from the Visayas and Luzon arrived in Bataraza. Here, they found a coastal plain dominated by seafaring and mobile communities of Muslim Tausug, Jama Mapun, and Samal, and scattered Chinese and European merchants.[1] These newly arrived Christian settlers peaceably accommodated Muslim authority in the region and entered into largely reciprocal trade relations with indigenous households. Early settlers often shared similar livelihoods and economic capabilities with the Pala'wan, with whom they traded as they initially established pioneering swiddens in lowland coastal forests to cultivate rice and root crops. During the initial years of settlement, many migrant families came to rely on Pala'wan households for subsistence as they struggled to establish their livelihoods. These commodity relations focused on migrant preference for rice as a staple—throughout the municipality, migrant settlers bartered medicine, coffee, and fish for rice and other staples from upland indigenous communities.

Oral histories conducted with indigenous peoples and migrant households throughout Bataraza suggest that the 1960s was associated with a substantial increase in paddy rice farming by migrant lowlanders.[2] This resulted in a shift in the flow of rice—now primarily flowing *from* the lowlands into the uplands—and a related change in economic relations between lowland Christian Filipinos and Pala'wan households. Before the arrival of significant numbers of Christian migrants following the Second World War, paddy rice farming was largely unpracticed by lowland Muslim communities in Bataraza, whose subsistence was partially based on the extraction of rice surpluses from upland Pala'wan *uma* plots.[3] Christian migrants from Luzon and the Visayas brought their traditions of wet rice farming, which were unfamiliar to Tausug Muslim settlers, whose livelihoods had begun to revolve around the production of copra. Land initially cleared as *kaingin* by Christian Filipino settlers was progressively transformed into permanent paddy fields and later became the basis of migrant tenurial claims in the emerging national bureaucracy on the island.

For these migrants, *kaingin* was viewed not as an enduring tradition or integral feature of rural life but as a stepping-stone toward the development of permanent and irrigated agriculture. As migrant paddy fields came to produce surpluses, indigenous households entered into enduring patron-client relationships with lowland farmers, who would advance Pala'wan people cash, rice, medicines, and coffee.[4] An elder Pala'wan woman described to me how Christian and Muslim lowlanders "taught" them how to drink coffee in the 1950s by offering them the drink while they labored. Though Pala'wan initially resisted working in lowland fields planting and harvest rice,[5] their debt obligations to Christian migrants and need for cash income eventually facilitated not only their participation in lowland labor but also their dispossession of the coastal plains.

As a result, Pala'wan households, which once resided on the current site of the *barangay* plaza in the lowlands, now almost exclusively reside on state forestland. Though the Muslim Tausug are frequently identified as the primary antagonists in Christian accounts of indigenous dispossession, it was not until the sustained movement of settlers from the Visayan Islands and Luzon to Bataraza that Pala'wan were largely forced into the uplands. An elder Pala'wan man neatly described the process: "The Christians first look for the land [from us]. They rented our land and loaned us money so they can plant their crops there. But because we don't have the money [to repay the loan] we eventually must hand over our land to the migrant. That's why the *bisaya* [Christian lowlanders] here owns a lot of land, because they have the money."

The period also marks the creation of Bataraza as an independent municipality (1964) and the rise of a more prominent lowland government in the area. While previously the area was nominally part of the municipality of Brooke's Point (1951–63), neither the municipal nor the national governments had the effective capacity to administer the territory. As the bureaucratic infrastructure of municipal government was established and former Tausug *datus* were incorporated into the new democratic social order, the social and economic dislocation of Pala'wan households emerged against new forest management practices that saw the uplands increasingly well-defined as a space to be actively managed. The landscapes of southern Palawan became not simply a source of subsistence for rent-seeking lowland polities but an object of governance for the Philippine state. Though these processes had been at work for centuries elsewhere in the archipelago, it was only in the mid-twentieth century that the Philippine national government had the capacity to map the uplands of southern Palawan through cadastral demarcation and subsequent enforcement of boundaries between the forested public domain and privately titled land.

These land categories were drawn from long colonial histories connected to agricultural and commercial imperatives that reified lowland paddy rice as productive and modern and understood *kaingin* as a backward and destructive practice. While the Spanish regime asserted their dominion over the archipelago through what has become known as the "Regalian Doctrine,"[6] it was the inability of successive colonial governments to effectively control the rugged, mountainous terrain that came to frame the "uplands" as a distinct frontier zone (Gibson, 1986, pp. 15–16; Scott, 1974). However, following the transfer to American administration in 1898, the assumptions of the Spanish legal fictions were grasped and remodeled on US public land and resource law in order to claim 92.3 percent of the archipelago as "public domain" (Lynch, 1986, pp. 270–71). Thus, when the newly independent national Philippine government was established in 1946, it inherited the mandate to classify, manage, and economically exploit the huge areas of forestlands as public domain—albeit ultimately for the benefit of a small oligarchic elite.[7]

As the American colonial administration (and later the Philippine national government) established bureaucratic infrastructure on Palawan Island in the early twentieth century, the uplands of southern Palawan were progressively delineated by the Bureau of Forestry in accordance with the preservation of heavily forested areas for future commercial exploitation. Areas deemed to have significant timber reserves were classified as forestland, while those more valuable for agriculture were released as alienable and disposable lands.[8] However, as elsewhere in the Philippines, the practical politics of releasing lands for legitimate agricultural development continued to support the ambition of wealthier lowland communities. These demarcations physically crafted a forested "uplands" in line with emerging economic disparities between Pala'wan and lowland migrants by releasing land on the coastal plains for agricultural development—thereby legitimating existing migrant paddy rice fields and coconut plantations. In contrast, Pala'wan households were excluded from the legal instruments of tenure by wealth, language,[9] and a criminal subsistence base (*kaingin*) and increasingly relegated to the forested uplands now claimed by the state as public domain because of land appropriation by migrant settlers.[10]

This division meant that the uplands, where indigenous people resided, had become an object of state territorial imperatives distinct from the private development of coastal plains (in theory if not in practice, as the ability of forestry officials in Brooke's Point to enforce state ownership or control over the forests of southern Palawan remained limited). Nevertheless, as the Forest Act of 1904 (No. 1148) (the legal basis of forestry policy until the 1970s) and other forestry laws explicitly criminalized *kaingin*, Pala'wan households in Inogbong and beyond were subject to sporadic enforcement of antiswidden measures following the Second World War. In discussion with me, an elder Pala'wan man reflected on economic motives during the 1960s–70s of forestry officials, who focused on preserving commercially valuable tree species while extracting fines from indigenous households: "They come here and check the posts of the house to see if they are *ipil* tree [*Intsia bijuga*].... If it is *ipil* you need to pay one hundred to two hundred pesos per post. I said, 'Just wait, but if you cannot wait I will use my blowgun on you!' and they were scared.... They told us that [*kaingin* is forbidden] because if we do that, we either we pay the fine or we will be imprisoned."

This division between coastal plains and hilly hinterland of the province, structured through the unequal economic relations between lowland Muslim and Christian communities and Pala'wan, would become elaborated upon and reified in the Revised Forestry Code of 1975 (Presidential Decree No. 705). PD 705 stipulated a systematic means of classifying public domain and appropriate land use based on slope and vegetation cover. This legislation considered all lands above 18 percent slope or containing significant forest cover as part of the public domain. Increasingly, this mandate articulated the need to distinguish public from private domain in terms of environmental protection—and, in particular, the role of human activity in soil erosion (Cramb, 2000)—rather than in purely economic terms.

While indigenous households in Inogbong had previously resided on and accessed a much wider section of the coastal plain, by the 1960s frontier settlement processes and the local politics of land tenure had seen Pala'wan households become primarily relegated to the state-owned forestlands of the *barangay*. These encounters between indigenous peoples and diverse state and nonstate actors in Inogbong produced a sharp sociospatial divide between an indigenous upland and a migrant-dominated coastal plain that came to have considerable implications as the capacity to administer and govern forestlands grew over time in the Philippines.

COMMUNITY-BASED FOREST MANAGEMENT AND CHANGING INDIGENOUS LIVELIHOODS (1980S–2002)

The relegation of many indigenous groups on Palawan Island to what had become classified as state forestlands meant that Pala'wan people in Inogbong would be targeted by new and widespread community-based approaches to managing forest resources in the Philippines during the 1980s–90s. While these approaches were seen to represent a break with harsh anti-*kaingin* policies of the 1960–70s, in which upland farmers were harassed, fined, or expelled from state forestland, new forest management programs on Palawan Island continued to draw on the enduring preconceptions of swidden agriculture to inform program design and practical interventions into the lives of indigenous peoples.

In 1995, the then-recently created Palawan Council for Sustainable Development (PCSD) began a seven-year catchment-based upland development and conservation project—funded by the European Union—targeting eleven *barangays* on Palawan. The Palawan Tropical Forestry Protection Programme (PTFPP, or simply "Tropical" to indigenous households in Inogbong) was enacted to support the PCSD's Strategic Environmental Plan (SEP) for Palawan. As a statutory land use management plan specific to the province of Palawan, the SEP formed the basis for PTFPP policy and project implementation. The PTFPP, however, was not designed in a vacuum. Previous waves of technical classification and production of knowledge regarding people and forests on Palawan strongly influenced the project's design and delivery. The endeavor to formulate a statutory land use management plan for the totality of the province was drawn from the precursor program, the Palawan Integrated Development Project (PIADP Phase I—funded by the then European Economic Community and Asian Development Bank). The PIADP (1982–90) aimed to economically develop and "arrest environmental degradation" on Palawan through, in particular, forest stabilization.[11] PIADP efforts to "stabilize" (that is, sedentarize) indigenous land use focused explicitly on swidden practices (Sandalo, 1994). Agroforestry interventions served as the technical means of "persuading tribal minorities to change from shifting cultivation to settled cultivation with annual and perennial crops" (Asian Development Bank, 1991, p. 4). This persuasive effort focused on overcoming indigenous communities' "basic lack of trust" in the government through a heavy reliance on foreign consultants and model farms in upland areas as a means to demonstrate the viability and economic success of agroforestry techniques to skeptical indigenous farmers (Asian Development Bank, 2002, p. 4). Several project sites in the south of the island promoted the integration of cashew, coffee, cacao, mango, and other cash crops into indigenous swidden fields through alley cropping and contour farming agroforestry techniques (Cramb, 2000).

While widespread throughout Palawan Island, the PIADP had few projects in the municipality of Bataraza and little presence in Inogbong beyond a seedling nursery and associated environmental planning studies. However, the PIADP facilitated the creation of a comprehensive land use plan for Palawan Island known as the Integrated Environmental Program (IEP), which would have far-reaching consequences for future community-based project design. A British environmental consultancy, with limited experience in the Philippines or knowledge of indigenous lifeways, devised technical subclassification of state forestlands based on vegetation type and elevation, which allowed for the development of commensurate approved land use activities (Hunting Technical Services Limited, 1985). In these planning documents, the management of land use types focused on associations between altitude, people, and agriculture that stipulated paddy rice for the coastal plains and (sedentary and constrained) agroforestry and subsistence production for the foothills. Ideally, these visions entailed no human use or occupation of the forest interior, where considerable numbers of indigenous people live.

When the SEP was legislated by the Philippine congress in 1992, the technical knowledge of people and environment assembled by the PIADP formed the basis of the proposed Environmentally Critical Areas Network (ECAN), which defined "a graded system of protection and development control in the province of Palawan" (Republic Act No. 7611). Like the IEP before it, the ECAN zoning scheme divided the entire island into differentiated resource use areas that stipulated an increasingly restrictive set of activities along the coastal-mountain gradient, culminating at the highest elevations in an untouchable "core zone," in which effectively all human residence and economic activity are forbidden. This basic zoning system was thus a bold act in rendering the landscape legible, technical, and manageable, infused with modern ideals of fixity, productivity, and surplus. In time, indigenous people's continued residence in the buffer zones was seen to be contingent on the adoption of sedentary agroforestry practices.

The PTFPP's primary objective was to reshape the disorderly reality of upland social life and forest ecology to conform with the graduated zoning of the ECAN. Like previous social forestry programs on Palawan, the PTFPP relied heavily on foreign consultants who drew on research from the then-named International Center for Research in Agroforestry and other development agencies regarding sustainable agriculture to formulate policy for indigenous livelihoods. Informed by widely circulating Edenic narratives of shifting agriculture, in which once-sustainable practices had become environmentally destructive due to changing cultural values and population densities (O'Brien, 2002), policy documentation positioned indigenous livelihood activities as the primary cause of deforestation and environmental degradation. Stabilizing upland livelihoods therefore became the core objective of the project (rather than curbing still-pervasive illegal logging). In building on earlier social forestry efforts, the PTFPP operated through a focus on livelihoods as the primary object of governance that could achieve both implicitly and explicitly spatialized objectives. In line with ECAN zoning, alternative livelihood planning and the delivery of social services on state-owned forestlands envisioned a reordered mountain landscape. This planning stipulated paddy rice for the coastal plains, rain-fed agroforestry production in an indigenous "traditional use zone," and ideally, no human use or occupation further in the forest interior (PTFPP, 2002a).

Undergirding this spatialization was the ECAN's "graded system of control," which mapped an idealized pattern of land use onto Inogbong. This zoning suggested that population density and livelihood activity should be progressively restricted, culminating in a "core zone." On the steeper slopes of the "core" and "controlled use" zones that supported a substantial number of households and their livelihood practices, PTFPP policy documentation in Inogbong stipulated either the complete absence of activity or nondestructive farming (*"[Hin]di mapanirang pagsasak"*) or no change in existing land use (*"Hindi pahihintulutan ang pagpalit-gamit sa lupa"*), respectively (PTFPP, 2002b). In other words, to be compliant with these guidelines, indigenous people would need to sedentarize their existing swidden fields through commercial agroforestry.[12]

Project documentation advised field staff to instruct farmers to "always use secondary growth areas, never cut primary forest" or "to reduce erosion and maintain soil fertility use flat or gently sloping land." Moving people away from sensitive areas with less mobile livelihoods could "assist rural communities to intensify and diversify their agriculture whilst at the same time safeguarding the forest" (Craggs, 1998). To address concerns regarding loss of income or subsistence production in areas where agricultural activity was to be limited, project officers instructed many households in restricted areas to engage in handicraft production for piecemeal sale in local markets. In contrast, households residing in the more gently sloping "traditional use zone" were supported in intensifying agricultural practices for market production. For example, fruit crops with (then) high commercial value were dispersed, and households that worked closely with PTFPP field officers were able to amass large plantations of fruit trees with varying levels of success.[13]

The *barangay* of Inogbong was targeted not because of its unique social or ecological characteristics but because of political ease. PCSD officials originally selected the forested areas of Barangay Marangas, also the location of Bataraza town proper. However, simply meeting the *barangay* captain proved difficult. A local PCSD official, who was involved in the planning and implementation of the project, described to me in detail the evasive strategies of the captain. Whenever the PTFPP team would call on his house to discuss setting up the project, the captain was "sick" or "out." Flatly saying no to anything in the context of Philippine sociality is a fraught negotiating technique, and it risks insulting and alienating potentially powerful benefactors. Instead, evasion is a common means of indirectly signaling refusal or dissatisfaction with an offered deal. While these kinds of projects offer potential benefits for local officials and their constituents, they are also accompanied by heightened scrutiny. At the very least, the PTFPP's entry into the area would require the captain to perform a considerably larger amount of work. After several weeks of effective evasion, project officials turned to the neighboring *barangay* of Inogbong, where the captain was enthusiastic.

As PTFPP project officers began working in Inogbong during 1996–2002, ECAN zoning regulations guided the delivery and promotion of nonswidden livelihood activities. As in other project areas, the core focus of the PTFPP in Inogbong was the provision of alternative livelihoods to fix people in place as a means to preserve forest cover, rather than efforts to police the complex boundaries of resource use zones. While training and materials for a range of livelihood activities were made available to indigenous households, the most sustained and widespread activity of project staff was the promotion of commercially oriented agroforestry as a means to curb the clearing and burning of forests for swidden. In contrast to the alley cropping promoted by earlier forestry programs in the 1980s, these agroforestry recommendations involved economically enriched fallows that encouraged the planting of high-value tree crops throughout the *barangay*. Large nurseries were established to facilitate the distribution of seedlings to indigenous households, and Pala'wan men and women were given cash and other material incentives such as clothes, food, or kitchen utensils in exchange for maintaining and planting perennial tree crops (Hobbes, 2000, p. 64). In individual household fallows, this meant establishing putatively high-value crops such as cashew, *kalamansi* lime, mango, *lakatan* bananas, or commercial rattan varieties that could provision for household subsistence through sale to local or provincial markets. The aim of "enriching" fallows and forests with commercial tree crops and rattan was to provide economic incentives against forest clearing for swiddens by making it more profitable to keep trees standing. Policy documents suggested that promoting this form of agroforestry to indigenous households "not only lessens forest dependency but also encourages participation in protecting forests" (PTFPP, 2002b, p. 28). In other words, under ideal enrichment scenarios, no guards would be necessary to protect trees, as Pala'wan themselves would be incentivized toward their protection as households progressively enriched more and more sections of the forest.

Policy documents did not imply that this spatial and economic logic of agroforestry was naturally desirable for Pala'wan households. Instead, investment in agroforestry solutions would need to be inculcated through the active labor of project staff in foregrounding connections between tree-cropping practices and the economic benefits and moral framings of forest protection. The primary tool for instilling these values was a widespread information dissemination campaign organized by field staff. This involved the distribution of educational pamphlets, enrollment in environmental seminars, visits to successful agroforestry areas elsewhere on the island, and generator-powered environmental film showings directly in villages. These efforts communicated and reinforced what constituted poor environmental behavior, such as cutting down large trees for *kaingin* or extracting forest products or wildlife (PTFPP, 2002b). The consequences of these activities were often linked directly to familial and community catastrophe, effectively moralizing environmental transgressions. For example, a PTFPP education booklet entitled *Kalikasan at kabuhayan*, or "Nature and Livelihood," tells the story of a rural community whose environmentally destructive livelihood practices result in flash flooding that washes away relatives and houses (PTFPP, 1997).

These activities sought to broadly engage all community members. However, project staff came to focus on building common investments in agroforestry within a core group of *panglimas*, who were framed in policy documentation as powerful brokers who could locally "govern" the implementation of conservation solutions (PTFPP, 2002b). While PTFPP staff emphasized their role as facilitators of new livelihoods rather than enforcers of forest policy, the ongoing monitoring of indigenous livelihood activities by field officers and deputized Pala'wan "upland development leaders," who resided in the foothill areas of the *barangay*, meant that many PTFPP projects in the area were viewed with suspicion by households in the forest interior (Hobbes, 2000). This was seen as a more effective alternative to relying on understaffed and ineffective local enforcement efforts from the Brooke's Point municipal government and *barangay* officials. A former *barangay* captain explained to me during an interview that "the PTFPP talked to the *panglimas* to oversee in their areas. They are our partners in the monitoring." One material manifestation of this plan was the construction of a series of community meeting halls in the uplands, which took their name and design cues from customary Pala'wan ceremonial houses (*kelang benwa*, or big house). These buildings were envisioned to be centers of local information distribution for these newly trained and empowered local leaders, who could broker a common reality of sustainable farming.

In a striking example of this effort to co-opt local political institutions, the PTFPP-established radio station proved crucial in disseminating information regarding the program objectives and appropriate environmental behavior. Radios were distributed to upland households under the "plant a tree, get a radio program," in which Pala'wan families were rewarded (and so incentivized) with radios for maintaining up to two hundred tree seedlings, which would then be distributed to other households or planted in reforestation activities. A former PTFPP project officer bluntly explained to me, "Well, it's like this, okay, all the livelihood programs are attached to the environmental protection. You protect the forest, you conserve the forest, we will give you this." In turn, these radios could disseminate information quickly across the mountainous uplands, ensuring that facts and ideas moved with great speed across the irregular topography that constrained project officers. A radio program entitled "The Hour of Indigenous People" (Oras ng mga Katatubo) was organized by PTFPP staff but featured compliant *panglimas* giving environmental lectures that encouraged fellow Pala'wan to plant trees within their plots (Hobbes, 2000, pp. 56, 66). The program foregrounded the economic and moral value of tree planting, with a prominent indigenous leader exclaiming to other households through the program that "if one is industrious, plants many trees and invests and time and effort in farming, there is no need to be hungry anymore" (p. 65).

Agroforestry interventions and disciplining of specific activities were essential, but not sufficient, conditions for achieving conservation objectives. In line with prevailing trajectories in Philippine social forestry, the PTFPP policy and practice were concerned not only with reshaping productive practices but also fundamentally reorganizing the lives of targeted populations through a discourse that closely connected environmental protection with human health, education, and residence patterns. Indeed, Inogbong itself was chosen because of its perceived deficiency, as one former PTFPP project official explained to me:

> Per the survey, the area is, in terms of illegal activities, also present.... There are more IPs [indigenous peoples], the *katatubo*, in the area, and they are more coercive, their communities. And in terms of health problems, they are very high at this time, and literacy is very low, and livelihood [is also low].... Of course the general aim of the project was to improve the quality of life of this people ... [to] improve the quality of life of this people through livelihood, education, and health. These are the three primary programs that we saw that really can improve the quality of life of this people.

This focus is borne out in the project's documentation. Improving health, for example, was explicitly linked to achieving environmental aims, and surveys conducted as part of project delivery sought to determine the prevalence of communicable diseases, such as malaria and tuberculosis, among indigenous households. In turn, project documentation identified the poor health of indigenous Pala'wan as a key dimension of their environmental destructiveness, as explained in the PTFPP-authored catchment management plan: "Health is wealth. Poor health means lesser wealth. Poor health requires needed income for medication. The natural resources are threatened as their exploitation provides the easiest means of acquiring extra income for medical treatment. Poor health also links to lesser productivity. This would mean lesser capability to work, lesser competence to lifer [sic] and lesser access to other resources other than the forest for living" (PTFPP, 2002a, p. 29).

In a concrete manifestation of efforts to fix people in place through the regulation of health, PTFPP and local government officials, in cooperation with the then-named German Agency for Technical Cooperation (GTZ), established a water supply system that could pipe water from two deep well pumps to the center of three upland hamlets (PTFPP, 2002a). As with many of the program's projects, the benefits were concentrated in areas deemed to be already degraded by human occupation, thereby encouraging residence away from environmentally sensitive areas identified under the ECAN. Similarly, education was identified in project documentation as a key focus, insofar as it increased environmental awareness: "Proper education is empirical [sic] for the community to understand the value of the catchment, hence, make them move towards its protection. As long as the dependents [sic] have grasped and have full understanding of [what] environment and resource management is all about, there's no more false hopes that the vision of having a protected forest will not carry on" (PTFPP, 2002a, p. 29).

This line of thinking underlay the establishment of a Tribal Learning Center located in the center of the *barangay*, which offered, at the time, free education for indigenous children in addition to adult learners. Alongside providing Pala'wan people with general numeracy and literacy skills, the Tribal Learning Center sought to explicitly convey environmental messages and promote a spatial concentration of indigenous residences. A report assessing the impacts of Tribal Learning Centers (TLCs) from 2002 highlights how nonenvironmental initiatives attempted to combine the interests of forest protection with the delivery of social services: "The establishment of the TLCs also encouraged IPs [indigenous people] living deep in the forest to relocate to the lower slopes. . . . The TLC was a venue for gathering of thoughts. Coordination of Panglimas with one another became more frequent. It has somehow captured the interest of the upland communities to venture in other means of livelihood without compromising their indigenous culture and tradition" (PTFPP, 2002a).

The community-based forest management efforts of the 1990s–2000s, under the SEP's ECAN zoning, were implemented through a series of strongly spatialized interventions in the state forestlands of Inogbong, which aimed to draw lives and livelihoods to what were seen to be more degraded areas and fix indigenous peoples in place. Not only were material resources, political power, and livelihood support directed primarily to the foothills of the *barangay* but swidden practices were also actively targeted in areas deemed ecologically sensitive, yet inhabited by substantial numbers of Pala'wan households. PTFPP project documentation presents a bold vision of transformation, but both the project's delivery and its afterlife highlight the unruly friction of governing people and places. Not mentioned in policy documents are the tactics used by even willing indigenous households to navigate the project's demands. One project officer emphasized in a lengthy review of the PTFPP's activities in Inogbong that when indigenous people were asked to perform free labor, such as moving heavy generators and project construction material, "they have so many excuses, like 'I'm busy,' 'I have work,' but when I said, 'Okay, find someone who can bring this, and I will pay them'—everybody was not busy anymore."

DEEP ROOTS: THE AFTERLIFE OF FORESTRY INTERVENTIONS (2003–2012)

As PTFPP funding ended in 2002, most of the technical and symbolic infrastructure (nurseries, project staff visits, radio programs, meeting halls, and the provision of incentives) that sought to rework indigenous livelihoods was dismantled, repurposed, or abandoned. The meeting halls constructed by the project, now largely dilapidated and rarely used, served as convenient places to conduct interviews or share meals. However, despite the failure of the project staff to secure further funding for operations and in the absence of any other effective conservation initiatives as of 2012, discourses surrounding livelihood change and decision making among many indigenous households continued to be made in reference to the agroforestry ideals and antiswidden arguments promoted by PTFPP officials some ten years earlier.

However, rather than representing a perfect continuation of the project's aims and aspirations, elements of the PTFPP's rhetoric and material effects operate through the lives of indigenous people in Inogbong in the absence of project staff. As part of seeking to understand upland livelihoods, alongside the typical ethnographic tools of participant observation and in-depth interviews, I surveyed Pala'wan households in each upland hamlet over 2011–12 about the kinds of activities indigenous peoples engaged in to provision for familial needs and about future economic aspirations. The results of these surveys suggested, mirroring local discourses, that differences in livelihood had emerged between indigenous households living further into the steeper and more forested interior of the *barangay* (roughly aligning with the ECAN's "core" and "restricted use" zones), and those in the flatter and less forested foothills (identified as the "controlled use" zone) across a range of markers of wealth. Most significantly, though few households had abandoned swidden agriculture altogether, households in the foothills of the *barangay* find themselves entering into market-based and more commercially lucrative activities such as irrigated paddy rice farming. These households are more likely to produce surpluses of cash crops (such as *lakatan*-variety bananas or *kalamansi* limes distributed by the PTFPP) or paddy rice or engage in income-generating activities such as daily waged labor (*arawan*) and reinvest in productivity-enhancing capital such as pesticides and carabao. For those farthest from lowland markets, swidden production remains the focal point of household life and activity. These households engage less in waged labor and rely more heavily on poorly remunerated activities such as handicraft production or the extraction and sale of *almaciga* (the resin of *Agathis philippinensis*, also known as Manila copal) and other forest products. Though there is considerable variation in household livelihoods within these broadly construed divisions, these differences are meaningful in the negotiation of everyday livelihood activities and social life. Why, then, have so many Pala'wan people moved away from swidden cultivation into the fixed kind of livelihoods envisioned by the PTFPP and other state actors, and why does this vary so significantly between different indigenous households across the gradient of the *barangay*?

The arrival of the PTFPP and the mapping of the *barangay* converged with a critical period in the late 1980s and 1990s that many Pala'wan now frame as a significant point of decline in swidden rice production. Survey data from Pala'wan households indicate that this perception of declining swidden productivity continues into the present (table 2.1). The decline in quantity and quality of rice, whether real or perceived, figures prominently in household narratives of diversification since the early 1990s and provides, at least in contemporary discourse, a key justification for the transformation or relative diminishment of swidden within larger livelihood aspirations. One Pala'wan man, reflecting on the period, linked the decline in the

TABLE 2.1 Self-reported rice yields in Pala'wan swidden fields, 1995 and 2010

	1995 ($N = 43$)	2010 ($N = 56$)
Average size of main *uma* plot (ha)	0.71 ha	0.68 ha
Average amount of rice planted (kg)	29.19 kg	30.20 kg
Average amount of rice harvested (kg)	239.00 kg	181.85 kg
Average rice yield per hectare (kg/ha)	336.00 kg	267.42 kg

Note: Data are drawn from a survey of sixty-three Pala'wan households across the gradient of the *barangay*. The number of responses used reflects whether or not the household made a swidden field during that year. The discrepancy between 1995 and 2010 therefore reflects younger or newly established households, whose members, while making a plot in 2010, may have been dependents within their parents' household in 1995. A yearly harvest includes rice from the first and second harvest, but not the smaller subsequent harvests that occur.

quantity of rice produced in swidden fields to the decline in quality, suggesting that that by 1995 "rice was no longer fragrant" (*kaya mebungu paray*). Another emic marker of decline is the disuse of rice granaries that are commonplace near swidden fields. Many Pala'wan suggest that in the time of their parents, such storage bins were overfilled with swidden rice but now sit disused. There is little point in filling such containers with the very small amount of recently harvested rice.

In response to the perception of declining yields, by the time of the PTF-PP's arrival in the *barangay* in 1995, some indigenous households in the uplands with access to flat land and readily available water had already begun experimenting in paddy rice production. Oral histories conducted with these pioneering Pala'wan paddy-farming households revealed that their pathways were facilitated by seasonal wage labor in lowland migrant fields. Not only did this instruct Pala'wan people in the pragmatics of irrigated rice production necessary to start crafting their own fields (initially by hand), but it sometimes also provided the capital to intensify production through purchasing or renting carabao or inputs such as insecticide or fertilizer. However, the ability to generate cash income or produce rice surpluses through paddy farming was fractured broadly along geographic lines. Wage labor is most accessible to communities in the foothills,[14] steeper slopes in the forest interior prohibit clearing fields for paddy rice of any significant size, and long distances over rough terrain inhibit the transportation of significant quantities of copra or other cash crops, another key source of income.

Alongside the perceived decline in swidden productivity, there were also important "pull" factors that motivated Pala'wan people to engage in new livelihood activities. Trajectories of change have seen Pala'wan women and men increasingly integrated into the lowland markets and social economies. Changing household requirements since the mass migrations of the 1950s are a key factor in this need to engage with markets to sustain households. Lowland commodities such as flavoring packets, coffee, and cane sugar—in addition to school fees, the regular purchase of clothing in the municipal markets, and the provision of rice in every meal—are now seen as necessities in the basic constitution and reproduction of indigenous households. These factors, alongside the desire for productive assets such as ploughs and carabao, mean that Pala'wan men and women see benefit in directing their energy to activities such as wage laboring and paddy rice farming, which can provide access not only to cash income but also, in the case of paddy rice farming, to *utang*, or credit relations.

This transition has continued since the start of the PTFPP in 1995, and the number of households engaged in irrigated rice farming has steadily increased in response to perceptions of declining swidden rice yields. For example, a Pala'wan man in the foothills who has now abandoned *uma* production entirely discussed his preference for paddy rice over swidden: "In *kaingin* it is very rare to have a good harvest, but paddy rice farming, after you have planted, you just wait for the harvest, and it has minimal maintenance. You just spray the herbicide and insecticide. So that is one of the advantages."

What role did the PTFPP play in promoting or accelerating these uneven transformations in the *barangay*? Rather than continuing uncontested, PTFPP ideals and practices have been partly taken up and refracted through the aspirations of indigenous people in the absence of any PTFPP project staff—particularly indigenous leaders, who could continue to broker ongoing interventions in the area. In the process, Pala'wan people foreground some elements of older rhetoric and discard others in an uneven accretion over time. This "mutation" of governance ideals helps explain the lingering, though transformed, impact of PTFPP material interventions and rhetoric long after the project's completion. A good example of this process can be seen in the integration of indigenous leaders into networks of governance through which forest management projects are brokered and delivered. By aligning personal livelihood ambitions with state agricultural and environmental priorities, emerging political elites in the foothills have been able to redirect the flow of benefits from agricultural and conservation projects via municipal and national state agencies toward their own household networks. While occasionally such transitions from swidden to paddy rice are developed independently, as many oral histories of change during the 1980s indicate, more frequently they are enabled by relations with representatives of state agencies and lowland paddy farmers. Pala'wan households who now cultivate productive paddy fields have frequently developed social networks that facilitate making loans or receiving aid packages to obtain capital inputs, or simply incurring debt, with neighboring migrant farmers for household rice consumption. For example, new paddy seed varieties are frequently distributed by the Department of Agriculture (DA) in order to boost productivity. In contrast, the DA encourages farmers further in the interior to sedentarize their swiddens and transform their cropping from *kaingin* to "upland," shorthand for plowed but rain-fed rice farming in the uplands.

In addition to issues of elite capture, the spatialized distribution of more lucrative livelihood activities allowed households in the foothills who grasped PTFPP agroforestry initiatives to develop sources of cash income and supplement declining swidden production. Some households in the foothills of the *barangay* have therefore come to specialize their household economic activity in irrigated paddy rice, commercial agroforestry, or wage laboring to support household needs. Engaged in a broad range of livelihood activities that supported cash income, many households that directly or indirectly benefited from PTFPP projects and new economic opportunities now de-emphasize production in swidden by planting fewer varieties of rice and other crops, shifting plot location less frequently, dedicating less effort toward weeding and guarding, and clearing in only minor forest regrowth. As a result, swidden plots see poorer production relative to those further in the forest interior (table 2.2). One striking visual example of this divergence

TABLE 2.2 Swidden productivity among Pala'wan households in the foothills and forest interior, 2010

	FOOTHILLS ($N = 34$)	FOREST INTERIOR ($N = 29$)
Average size of main *uma* plot (ha)	0.69 ha	0.66 ha
Average amount of rice harvested (kg)	152.47 kg	213.40 kg
Average rice yield per hectare (kg/ha)	221.97 kg	323.33 kg

can be seen in the swidden-mapping exercises in my livelihood surveys. In the picture of two fields made by households of roughly comparable demographic composition, there is a clear distinction in agrodiversity between the household residing further in the forest interior and the household relying on paddy farming (figure 2.1). Furthermore, households who focus away from swidden often recrop the same land for more years and, rather than abandoning old swidden fields to fallow, convert them into agroforestry plots. While swidden remains a vital part of household provisioning, its role within broader livelihood strategies has moved from center to periphery as socially and spiritually important rice supplies are sought from either lowland markets or paddy fields.

In contrast, distanced both spatially and socially from lowland centers of power, households in the forest interior find their ability to negotiate declining swidden production and increasing household demands diminished. Though the ability to engage in activities that provide cash income is

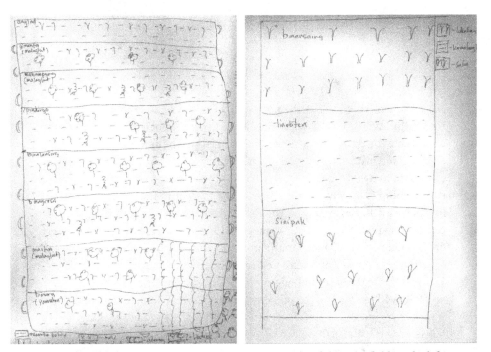

FIG. 2.1   Crop maps comparing two Pala'wan swidden fields. The field on the left is from a household in the upper catchment; on the right, from a household that relies far more heavily on paddy rice. Varieties of rice are labeled in horizontal divisions

ROOTED IN PLACE   81

FIG. 2.2
A Pala'wan man plowing his paddy field in the uplands

not excluded from the forest interior, it is limited relative to households in the foothills. Due to a steep mountain gradient, land becomes progressively unsuitable for paddy farming and agroforestry. Though many households experiment with paddy production, it is mostly on a scale that is minor when compared to fields closer to the lowlands—men who make such fields (paddy rice farming is largely a male undertaking in upland Inogbong[15]) are often inexperienced at overcoming the difficulties associated with wet rice production in steep terrain. For example, during the course of my fieldwork, I observed a young Pala'wan man's upland paddy field decline in size as the soil around it eroded and rendered the plot unworkable (figure 2.2).

In response to these challenges, marginal households often "lean harder" on their swidden plots for basic subsistence, the provision of rice, and dietary diversity, by investing more effort in the process of swidden making. These households must also therefore negotiate the imprecise categories and zones that surround tree clearing and are often arbitrarily enforced by lowland state officials—for example, the "controlled use zone" and "restricted use zone" of the ECAN have little meaning on the ground (and are subject to change), and they are operationalized by *barangays*,[16] municipal employees, and Pala'wan themselves as "big trees" to provide an intuitive guide to what is allowed where.

In addition to the unevenly distributed benefits and burdens of governance during the 1990s and early 2000s that have inflected diverging pathways of livelihood and relative wealth, the inculcation of environmental narratives and corresponding forms of community discipline continue to shape how, and where, indigenous peoples make their livelihoods. Despite a decade-long absence of PTFPP project staff or any other sustained presence, the production of conservation spaces continued through the ongoing reinvestment in the binding logic of tree-cropping solutions and the benefits of sedentary lifeways by some farmers themselves. Interviews with indigenous people over 2011–12 revealed the ongoing, and contested, promotion of sedentary lifestyles as morally and economically superior to mobile forms of swidden, often in reference to PTFPP environmental discourse in ways that converge with emerging forms of social difference.

Those best able to grasp the emergent political economy of conservation have, to some extent, internalized the environmental narratives of the state and civil society and use them to explain their own personal livelihood transformations that focus on nonswidden alternatives. For example, a Pala'wan man engaged in paddy rice farming and commercial agroforestry reflected on the legacy of the PTFPP through discourse that frames fixed forms of living in environmental and moral terms:

> When the PTFPP came, they made the "*bawal* [forbidden] issue" clear to us, and we learned to fight for ourselves. We reasoned, "How do we live, since we don't own a carabao?" But the truth is, [since] making *kaingin* is prohibited then, it's prohibited to make *kaingin* in the forest or mountain because it would make the water disappear and mountains will become bald. Our life had improved since the PTFPP came, and that's why I'm so thankful to them, because they taught us to stay in one place and do the planting and avoid transferring from place to place in search for good land. Our children now can go to school, and they told us not to be afraid of Muslims or Christians. They also told us that we should plant more trees in our *kaingin* and increase our number of crops. . . . Right now I'm training my children to do the *basakan* [paddy rice], because it's much better than *kaingin* in terms of harvest.

A *panglima* from a hamlet in the foothills reflected further on the sedentary logics of PTFPP rhetoric:

> Sometimes they said that even though you have *kaingin* and planted many crops and you transfer to another place, you will notice that the wild pigs will eat your crops, so where will you get now your food for daily consumption? So they told us not to transfer into another area to do *kaingin* because if you already cultivate one area and your plants are already big enough, if you will transfer, that means you will have to start again. So I also agree with them because, for example, it's time for harvest and you are in another area, your crops will just die, and you can't even get the harvest, and you have to start again, so they said there's no good future in that.

The ongoing reproduction of PTFPP rhetoric means that households in the forest interior live under conditions of uncertainty regarding the enforcement of environmental policy, as their livelihoods are officially criminalized through the island's environmental zoning and its partial enforcement. While the actual presence of local government representatives from the municipality or *barangay* are limited, stories of paid informers, state helicopters, past arrests, and individuals being personally summoned to the mayoral office are in constant circulation and contribute to a broader climate of doubt and fear surrounding upland cultivation that is effective in limiting clearing.[17] During one interview, I asked a Pala'wan man why he did not clear old, larger trees for swidden, despite the many benefits. The ongoing exchange between this man and the Pastor was indicative of the persistent yet nebulous impact of older enforcement strategies:

> **PALA'WAN MAN:** Because someone told me that it is forbidden to make a *kaingin* in the forest.
> **PASTOR:** It's forbidden. . . . Who told you that it's forbidden?
> **PALA'WAN MAN:** The PTFPP.
> **PASTOR:** Who else?
> **PALA'WAN MAN:** That's all I know.
> **WILL:** But the PTFPP is gone.
> **PALA'WAN MAN:** They still have a guard here.
> **PASTOR:** Who?
> **PALA'WAN MAN:** They are watching the chainsaw operators, and they come here and tell us that cutting down trees is forbidden and that

moving our *kaingin* is also forbidden. The PTFPP says to only stay in the area that we have previously burned.

This is not to say that the enforcement of environmental regulations is perfect; "criminal" activities take place at higher elevations—wildlife is extracted and forests are felled. Nevertheless, this uncertainty is understood to have a pronounced impact on households living in the forest interior, which are subject to the persuasive and normalizing impacts of rhetoric promulgated by neighbors and kin. For example, a male Pala'wan elder who migrated from the forest interior to the foothills noted, "The PTFPP used to tell us about that before. They explained that it's prohibited to collect *yantok* [rattan], catch birds or even wild boar or any animals, cutting down big trees. . . . I already knew it, and that's what I'm telling my companions about, but they do not heed my counsel. They answered me, saying, "What if our children starve? Will it [the PTFPP] feed them?" And about the *kaingin*, I heed the PTFPP's advice; I only plant rice in my paddy rice fields."

The livelihood practices of contemporary Pala'wan have become highly differentiated across the gradient of the *barangay* as households respond to local perceptions of agricultural decline. In 2011–12, productive capital and political leverage among indigenous households was concentrated in the foothills of the mountain range. This has enabled some of these lowland households to gain greater command of the land and labor of other Pala'wan in the catchment. Households in the forest interior, in contrast, are increasingly marginal to emerging economic opportunities. Many Pala'wan link this social differentiation to the policies of the PTFPP, and they legitimate or resist their socioeconomic position according to the language and moral logic that was been introduced to human-environment interactions by conservation projects and information dissemination campaigns.

This broad survey of Pala'wan livelihood discourse points clearly to a persistence of older governance ideas and suggests associations between emerging social difference and the ongoing influence of PTFPP rhetoric. The enduring, yet reconfigured, nature of PTFPP rhetoric is sustained through everyday personal ambitions and individual social acts of production and consumption. The following two intimate portraits of uneven but intersecting lives, the livelihoods and relationships between two *panglimas* and their household members, whom I interviewed and repeatedly visited throughout my fieldwork, give specificity to the ways in which projects of governance can come to have a lingering impact on place as routinized and everyday experiences of uneven power. These case studies of two particular households reflect broader, interrelated dynamics of resource access, power relations, and swidden transformation unfolding in the *barangay* between the forest interior and foothills. Panglima Burit and Nerma Labut, who reside in the foothills of the *barangay*, represent an indigenous household that has successfully navigated community-based forest management efforts to establish themselves as relatively wealthy indigenous people. In contrast, the household of Panglima Murta and Lina Tanduk has experienced livelihood interventions and local enforcement of forest policy in largely negative terms.

BENEFITING FROM FOREST MANAGEMENT: PANGLIMA BURIT AND NERMA LABUT

Burit Labut and Nerma Labut live in the hamlet of Saray with their two daughters. The matrilocal hamlet of Saray was focal point of the PTFPP's activities in the *barangay*. Burit and Nerma's eldest daughter, Narhelin, attends the Bataraza campus of the provincial Palawan State University, one of only two Pala'wan in the *barangay* to do so. Though Burit identifies as a Christian and attends the sporadic Baptist services held in the small upland churches, he practices important rituals that call on Pala'wan deities and powerful ancestor spirits.

Peminta Sandar, Nerma's father and former *panglima* of the hamlet of Saray, has been a considerable influence in the life of the family. Peminta was widely respected throughout the uplands as a political leader and ritual specialist. This knowledge has been passed down to Nerma, Burit, and their daughter Narhelin. Thanks to her education, Narhelin can deftly navigate discussions of both Pala'wan cosmology and ethnoastronomy and the matters of contemporary Filipino life. In addition to his ritual knowledge, Peminta was also considered one of the first members of Pala'wan households in the uplands to cultivate irrigated paddy rice. However, after Peminta's death in the mid-1990s, the paddy field was abandoned. Burit says that the old man visited him in a dream after the funeral and instructed him to continue making a paddy field in that area and encourage his wife's brothers to do likewise. He worked closely with them in the gentler slopes closer to the coastal plain to construct earthen dams and more effectively regulate irrigation schedules. Because of these efforts, he now maintains a half-hectare paddy field started in 1997, made on alienable and disposable land just outside of the state-owned forestland inherited by Nerma from Peminta.[18] However, the decision to cultivate paddy rice also had a more pragmatic impetus: "I've learned it from my father-in-law, because before, he used to work for a friend in a paddy field, then after some years, he decided to have his own field [in the early 1990s]. So, as I observed him, I learned some techniques, and I also observed that his financial status became stable. He has a good harvest compared to *kaingin*, which has so many difficulties, like pests and the variable weather. So I decided to have my own paddy field too."

Despite this, the couple also makes a swidden field each year of around half a hectare. In contrast to the irrigated rice field, making the swidden is a joint endeavor for the couple. In 2011 they planted three varieties of rice— *linapat* (glutinous), *sambirara*, and *anglad*—intercropped with cassava, sweet potato, corn, and a variety of vegetables that border the plot. As in other households, when invited to seminars or workshops that routinely occur on Palawan Island, Burit adamantly defends the right of indigenous people to practice swidden. However, while Burit leverages the right to cultivate in forested landscapes in public forums, other Pala'wan households suggest that he lectures them with rhetoric gleaned from the PTFPP and directs them not to open new sections of forest for cultivation and, instead, recultivate older fields. Many of his neighbors see him as the ongoing representative and deputized enforcer of the PTFPP's wishes. In practice, the family is increasingly less reliant on swidden for cash income or subsistence. As such, he and his wife have mostly sedentarized their swidden practices, preferring to make their *uma* in a location with limited regrowth but close to their house. In 2012 they began to experiment with sedentary ploughed fields at the behest of a Department of Agriculture initiative coordinated locally by Burit. Such experimentation, though aimed at poorer farmers, requires access to a carabao in order to successfully plow upland plots.

Nerma is far less politically active than her husband, who, in contrast, has taken a leading role in brokering forest management efforts in the *barangay*. During the PTFPP's involvement in the *barangay*, Burit, following Peminta's lead, was actively involved in disseminating environmental regulations and restrictions from ECAN zoning to other upland Pala'wan. Based on these connections established with field staff and members of other government departments during the PTFPP, Burit is often selected to participate in seminars, educational meetings of the Department of Environment and Natural Resources, the Palawan Council for Sustainable Development, and various NGOs in the provincial capital of Puerto Princesa. He reflected on the role of the PTFPP in facilitating ongoing opportunities: "If you look back to before PTFPP came, I am already here. I followed everything that the DENR said. Since there are many different [hamlets], every task that was asked of us, I made sure that we did it together. And they have seen that in Inogbong I will be the representative of the indigenous people, because I have done all that they asked for, from PTFPP until today."

These engagements have provided ongoing opportunities in terms of authority to allocate or organize project resources toward his household or kin networks. For example, Burit is the local contact of the Brooke's Point Department of Environment and Natural Resources office. Through this role, he helps facilitate projects in the *barangay* by organizing the mobilization of labor and access to land. In 2012, he worked with the *panglimas* of the hamlets of Papan and Paratuong in the foothills to establish rubber tree plantations under the DENR's National Greening Program. The DENR has established a nursery adjacent to his swidden field, and Burit coordinates laborers to construct the nursery and maintain the seedlings that will ultimately be planted on land he claims as his own.

However, perhaps as importantly, these engagements over time have also given him increased confidence in negotiating with lowland authorities, an experience that many Pala'wan find confrontational or uncomfortable. For example, though the *almaciga* tree concession was controlled for several decades by the powerful Rodriguez family, new licensing regulations stipulate that the "core zones" of the island (where *almaciga* trees are found) cannot be commercially exploited, except by indigenous people (PCSD Resolution No. 04-233, 2004). As a result of his increasing familiarity with environmental regulations and the state bureaucracy that manages them, Burit has been able to locally position himself as the de facto *almaciga* concession holder, by connecting to buyers in the provincial capital and partially fulfilling DENR licensing requirements. While this is endeavor is framed in terms of "alleviating the suffering" of his fellow Pala'wan, he maintains tight control over the flow of resources from the uplands and deploys his knowledge of and connections to local enforcement agencies to ensure that the *almaciga* is not sold to other buyers. For example, a Pala'wan man living in the upper portion of the catchment discussed his arrangement with the *panglima*: "But if he knows that you sell to someone, he will also get his percentage from that.... As a matter of fact, my eldest son sold to Burit fifty-five kilograms, and his companion has thirty-two kilograms. And if he knows that you are keeping some for yourself and you want to sell it to someone else, he will report you to the authorities and will accuse you for selling without a permit. If you can go to his storage house, I think the *bagtik* [*almaciga* resin] he has is almost six tons."

The hard work of the couple has not been in vain; their lives are tangibly different from many Pala'wan in neighboring hamlets. Their house is large and roofed with galvanized iron in place of nipa palm. Their kitchen sits in an adjacent hut and contains a salvaged metal sink connected to a

PTFPP-constructed water supply through an elaborate network of pipes. The interior is decorated with certificates of participation from workshops and projects attended by both Burit and Nerma. They own a television and petrol generator, which although infrequently used are rarities in the forested uplands. While these differences may seem slight in the context of larger inequalities between indigenous people and many lowland Filipinos, such distinctions are significant in the uplands.

BEARING THE BURDEN OF CONSERVATION: PANGLIMA MURTA AND LINA TANDUK

Efforts to manage forests on Palawan Island can also be understood by indigenous people to negatively impact their livelihoods and well-being by constraining swidden agriculture and other forms of forest resource use. Murta and Lina live with their five children in the hamlet of Manga-Manga, only a relatively short walk away from Burit and Nerma's house but within the "controlled use zone" during the tenure of the PTFPP and surrounded by significantly more forest cover. All five of their children reside within their household, and the two youngest daughters attend the Tribal Learning Center established by the PTFPP and now maintained by the government school system in the hamlet of Papan. Ailen and Jobilin walk between one and two hours to reach the school each day.

Though Murta sometimes labors in lowland fields, and Lina sporadically sells household brooms (*walis tambo*) in the town market, they have inconsistent sources of cash income. Without a steady cash income, they are unable to regularly purchase lowland paddy rice and highly desired condiments, such as MSG sachets and sugar. Their swidden is therefore their primary source not only of subsistence but also of dietary diversity. The couple thus place great significance on swidden; they plant a wide variety of rice and other crops and seek out mature secondary forest regrowth as an ideal site for clearing as a means of limiting pest and weed incursions. In 2011, Murta cleared an *uma* of approximately two hectares. The couple planted ten varieties of rice—*sambirara* (glutinous), *binere* (glutinous), *bandar, gandinga, langgung* (glutinous), *binura, binate, sinaligsig, inanut, gilingan*—intercropped with cassava, sweet potato, millet, sorghum, pigeon pea, yam, and taro. Despite the effort expended to facilitate swidden productivity each year, rice from their *uma* is never enough to supply each member of the household with rice each day. Though Murta is reputed to have the most productive swidden in the *barangay*, the household will usually rely on root crops (primarily cassava) many months prior to the next season's rice harvest. Like Burit and many others in the *barangay*, Murta believes his swidden has become steadily less productive over the past two decades.

In this context, the continued practice of agroforestry initiatives introduced by the PTFPP not only were a morally and environmentally approved way of utilizing forest spaces but also became a marker of productivity and social differentiation. In discussing livelihood change, Murta articulates a common position that connects success in agroforestry, in this case Burit's banana plantation, to participation in the PTFPP and ultimately economic success. As he later explained: "Even without *kaingin* we are able to plant *lakatan* bananas [a high-value variety]. As you can see with Panglima Burit, almost every day he harvests *lakatan* bananas. . . . Those who follow the PTFPP, of course they prosper, but those who did not follow the PTFPP have stayed the same. How can they prosper if they did not follow the teachings of PTFPP? But if they obey the PTFPP's explanation, they might prosper. For example, Panglima Burit can now send his children to school. Because of his involvement in the project."

Murta thinks that the increasing restriction on clearing has negatively impacted their ability to cultivate rice. He idealizes moving his swidden plot frequently, and sees rooting cultivation in one plot as the cause of low soil fertility and increasing deforestation, a sentiment common among households in the forest interior:

> For me, the reason why our harvest is declining is because we can't clear the large trees. Because you know if you don't clear in large trees, before you even *tugda* [dibble rice] there are many weeds, but if you clear in an area with large trees, your rice will become large in size. . . . If it is not forbidden, I would make a *kaingin* in the area with large trees, but since the *gobyerno* [government] won't allow us, we just stay [recultivate] in our old *kaingin*, but before it is not forbidden that's why we have a good harvest, but now we only cultivate in the cogon grass area . . . because we are not allowed to make *kaingin* so the result, the weeds grow there.

Murta knows that if he clears trees and is caught there may be significant repercussions. It is difficult to clear and burn trees in secret. He has heard that Panglima Burit and others who were close to the PTFPP field staff will report his illegal activities to the lowland *barangay* officials, and he believes that the mayor's auxiliary environmental protection force, Bataraza

Bantay Bayan, has hired informers among the indigenous community to monitor the clearing of trees. The DENR, other environmental agencies, and the municipal and state government are often conflated in discussions with indigenous people as the *gobyerno*, a term that may refer to one or more aspects of the Philippine state. Many households in Inogbong are therefore suspicious of any government activity or presence, as it is often closely associated with punitive control of forest resources. While Murta's wife is shier and cannot speak Tagalog well, Murta is openly critical of Burit and suggests that by associating himself with ongoing forest management efforts, he has "lost his culture." The criminality of clearing forest and declining rice yields make him, like many of his neighbors, consider alternative ways to provision for household consumption. While they are not excluded entirely, their distance and unfamiliarity with lowland markets limits their ability to engage in paddy rice, agroforestry, waged labor, or other endeavors and negotiate perceived swidden decline.

These descriptions of different household livelihood practices and histories serve to illustrate both the economic distances and the interdependencies that characterize social differentiation in Inogbong, broadly seen by many Pala'wan to be unfolding across the uplands of the *barangay*. For example, Burit has leveraged past involvement in environmental governance projects to gain disproportionate control over forest resources (primarily the *almaciga* trade) and direct the flow of employment from new state projects toward his own kin networks. These case studies also highlight how both discourses and specific practices surrounding swidden illustrate the impact of forest governance in diffused and deeply personal ways.

CONCLUSION

While the establishment of the Mount Mantalingahan Protected Landscape in 2009 might have suggested that the boldly defined boundaries of the protected area would become new sites for environmental action and enforcement, the MMPL is only the most recent in a long line of forest management and conservation measures in southern Palawan that give little attention, on the ground, to borders in the creation and contestation over space. The Palawan Tropical Forestry Protection Programme was a significant and comprehensive example of efforts to implement a community-based forest management project in the Philippines. While the PTFPP was designed to enforce complex ecological zones that now encompass Palawan Island, project design and delivery by staff focused instead on reconfiguring environmental subjectivities and bodily conduct (as movement through space) through promoting sedentary lives in particular locations.

Rather than fading into obscurity, the PTFPP remained a potent element of indigenous debates over how people should live in forested landscapes over the course of 2011 and 2012. The ways in which the strategies of persuasion and indirect mechanisms of rule achieve the desire of the state without shattering an individual's sense of autonomy can be thought of as "government at a distance" (Miller & Rose, 1990). "Distance" refers to both a form of spatiality and an absence of visible coercion, but the aftereffects of the PTFPP also speak to the temporal nature of efforts to shape people and place "at a distance." In Inogbong, the rhetoric and impacts of the PTFPP's livelihood interventions have mutated, been repurposed, and taken on a life of their own beyond the intentions of policy makers or long-since-departed development consultants. Social difference between indigenous households in Inogbong has become strongly spatialized across the mountainous gradient of the *barangay*, and it is often still explained as a result of the PTFPP's targeted interventions and restrictions. Politically active and wealthier households, in turn, often deploy the highly moralized rhetoric of the project to emphasize the benefits of sedentary, market-based living to recalcitrant kin and neighbors, many of whom continue to resist such transformative efforts.

Careful review of the lineage, implementation, and outcomes of the PTFPP reveals a more diffuse model of spatial production in conservation territories that takes place in the absence of strictly enforced borders and boundaries. More than this, however, the continued influence of the PTFPP a decade after its retreat from Inogbong demonstrates the ways in which upland spaces on Palawan, composed of diverse and changing indigenous livelihoods, continue to be produced as part of a contested process that frequently occurs in the physical absence of the state.

CHAPTER THREE

Insidious Vulnerabilities

THE vulnerability of indigenous peoples is a key component of global anthropogenic climate change narratives, and it underpins a host of developmental interventions and international projects of resource redistribution. Even within academic circles, the potency of this truism has meant that questions are rarely asked about what conceptually undergirds the calculation of indigenous innocence in the face of climate change.

In 2009, the indigenous advocacy group Survival International produced a global review of development aggression experiences. Survival International's report explains how the construction of hydroelectric dams, the expansion of cattle plantations, and the establishment of protected areas means that indigenous peoples are the group most affected by climate change (Survival International, 2009). Indigenous people, the report argues, "are on the frontline of climate change. Living in parts of the world where its impacts are greatest and depending largely, or exclusively, on the natural environment for their livelihoods, culture and lives, they are more vulnerable to climate change than anyone else on earth" (p. 1). Survival International's analysis examines how wider structural forces are culpable for producing poverty, dispossession, or marginalization that gives rise to vulnerability: colonialism, capitalism, or the development project. This is not a new perspective. The political ecologist Michael Watts, in his seminal and still-influential book, *Silent Violence* (1983, p. 19), suggests that understanding the precarious position of rural households and their uneven exposure to climate risks requires "a careful deconstruction of the social, political and economic structure of the society." As the deployment of terms such as "vulnerability" by Survival International and similar groups would suggest, this critical approach is also a politico-moral project. As human geographer Jesse Ribot (2014, p. 2) explains, "Most policy-oriented analysts of climate-related vulnerability shy away from historical political-economic analyses of causalities. They focus on identifying who is vulnerable rather than why. They seek indicators rather than explanation. This is no surprise. Causality is threatening. It implies responsibility, blame and liability." Determining the casual structures of vulnerability involves not just diagnostic evaluation of weakness or fragility but also unearthing what Ribot terms the "root causes." By charting the causal chains that lead to the marginalization or poverty of particular groups, this formulation of vulnerability offers the potential to calculate culpability amid a wider historical moral accounting.

Attention to historical political economic forces has constructed convincing narratives that shift blame away from small-holders toward larger and more abstracted forces (or their state or corporate manifestations). But what role do people's own explanations for vulnerability and practical responses to environmental pressure have within these frameworks? Is it possible to take seriously the intentions and actions of interlocutors—self-blame, acquiescence, or apathy—while maintaining the moral project of vulnerability? What does it mean for dominant conceptions of vulnerability when the marginal produce morally troubling explanations of environmental change and, seemingly, materially orchestrate their own fragility?

In regard to deepening capitalist relations on the Indonesian island of Sulawesi, anthropologist Tania Li (2014, p. 97) points to a troubling moment in agrarian transformations when Lauje highlanders responded to practices of enclosure "not by debating it but by recognizing and helping to enforce" capitalist relations between kin and neighbors. However, "rather than fault highlanders for what they did," Li points to the "insidious" nature of capitalist relations, which gradually eroded the choices of indigenous small-holders and increasingly bound them to a system of competition. This gradual nature of transformation in the Sulawesi highlands aligns with the notion of "slow violence" as processes that occur "gradually and out of sight, a violence of delayed destruction that is dispersed across time and space" (Nixon, 2011, p. 2). Attending to vulnerability as a project of slow or gradual violence can help capture less straightforward causal pathways that weave awkwardly through time and place to render indigenous peoples more or less exposed to changing weather patterns. However, as Li's example suggests, *insidious* social and environmental change entails more than spatial and temporal distancing. In this case, our collective scholarly attempts to narrativize suffering as part of a moral and analytical project of vulnerability are confounded by the seeming complicity of its victims. Working through these complexities, though perhaps not morally resolving them, requires an ethnographic focus on how these kinds of trajectories are positioned within struggles over place and the processes of subjectification that shape and constrain livelihood desires and decision making.

In 1997–98, Pala'wan people experienced a devastating El Niño drought. While this event has entered into popular and academic memory as a significant environmental catastrophe, recorded accounts from those who actually experienced its impacts are few.[1] Elderly and middle-aged women and men in Inogbong, those with established households and livelihoods in the mid-1990s, recall the drought as a time of subsistence crisis. Indigenous households experienced considerable hardship as agricultural productivity collapsed and cash income–generating activities failed. Pala'wan people were largely sustained not through the largesse of state emergency funds, but through reliance of emergency forest foods and a moral economy of food exchange that exists between households across the gradient of the Mount Mantalingahan range. However, despite potent memories of hunger and deprivation experienced during this time, these same farmers and their children now increasingly and willingly forgo swidden agriculture, residence in the deep forest interior, and "traditional life" more generally. As the move away from swidden intensifies, Pala'wan households are increasingly exposed to the impact of extreme drought as traditional subsistence safety nets used to cope with food shortage are diminished. Like the highlanders described by Li, they are seemingly complicit in their own vulnerability over time. Who, then, is to blame for the production of this acute vulnerability?

THE 1997–1998 EL NIÑO DROUGHT

Throughout Southeast Asia, shifting cultivators carefully scrutinize weather patterns in relation to agricultural activities. Deciding when (and what and how) to clear, burn, and plant requires consideration of a complex array of variables such as heat, wind, and precipitation, which vary considerably through time and space. Across the region, meteorological research suggests that seasonal rainfall patterns are largely determined by shifting monsoonal regimes. Annual cycles in any particular location, however, are contingent on local topographical variation resulting in pronounced or diminished seasonal distinctions and are subject to inter- and intra-annual variation. The El Niño–Southern Oscillation (ENSO) phenomenon heavily influences rainfall anomalies throughout Southeast Asia. ENSO refers to the connected shifts in sea surface temperature in the eastern Pacific Ocean from warm (El Niño) to cold (La Niña) phases—the atmospheric expression in these changes is associated with the pressure differential between Darwin and Tahiti (measured by the Southern Oscillation Index). The El Niño phase is characterized by an intense warming of the eastern Pacific and a decrease in atmospheric pressure in Tahiti and a corresponding increase in Darwin, which results in a weakening of Pacific trade winds and ultimately reduced rainfall in Southeast Asia. In La Niña phases, the eastern Pacific cools, which is associated with increased precipitation throughout Southeast Asia. Extreme manifestations of ENSO events in the region are accompanied by climatic disasters in the form of flooding and drought, with profound impacts on rural livelihoods.

While El Niño events are frequently associated with diminished rainfall, the 1997–98 drought is widely acknowledged as particularly dramatic. Throughout Southeast Asia, the impacts on rural livelihoods were severe. Global media reported heavily on extreme drought, food shortages, and forest fires. The *New York Times* dubbed the event "one of the most widespread man-made disasters the region has known" (Mydans, 1997). Within the Philippines, the event significantly reduced rainfall during the second half of 1997 and the first half of 1998, impacting two critical paddy rice cropping periods by shortening the length of the 1997 southwest monsoon and delaying its onset in the following year (Bensel, 2005). As a result, approximately 70 percent of the country experienced a severe drought, resulting in significant decline in rice and corn production, water shortages, and extensive forest fires (Hilario et al., 2009). The event disproportionately impacted the south of the archipelago (including Palawan Island), which is broadly seen to be most vulnerable to El Niño–induced drought. In southern Palawan, local climatic data indicate below-average rainfall during the usually wet southwest monsoon season, becoming particularly acute in the final quarter of 1997 and early 1998 before sharply transitioning to La Niña conditions at the end of the year and above-average rainfall.[2]

Because of the severity and devastation caused by the 1997–98 event, the term "El Niño" has, as elsewhere in the world, entered the common lexicon throughout the Philippines. Both Pala'wan and migrants in Inogbong use the term "El Niño" to identify the events of 1997–98 as distinct from other periods of drought, which are sometimes framed in Tagalog as *matinding init* (intense heat) and *sobrang init* (excessive heat), or in Pala'wan as *kelang bulag* (big heat). In oral histories conducted with indigenous families who experienced the full force of the 1997–98 event, many Pala'wan people emphatically suggest that there was no rain for seven to nine months, approximately between December 1996 and August 1997, with below-average rainfall continuing until 1998.

Because of the severity of the event, many informants would anchor their livelihood histories in relation to these experiences of drought. Pala'wan individuals who had established their households and livelihoods by 1997 suggested the impact of lower-than-expected rainfall on swidden production was uniquely devastating within living memory, matching the experiences of small-holders throughout the archipelago and Southeast Asia. The 1997–98 drought was compounded by below-average rainfall in 1996, and swidden crops planted in March to May of 1997 quickly died as substantial rain failed to arrive. While it is unsurprising that the lack of rain crippled climatically sensitive rice during sensitive cropping periods, many households recalled that drought conditions also stunted even hardy cassava and other staple root crops. Usually resilient banana and coconut trees also failed to grow. Intense heat damaged existing stocks of food, such as cassava in old swidden plots, which also became inedible (figure 3.1). Over the course of the drought, many indigenous households experienced a total crop failure within upland plots. One Pala'wan man recalled that the intensity of drought meant that not only did food become increasingly unavailable but water, especially in the foothills, also became scarce: "The big drought that happened before affected the whole of Palawan. Our problem was the food but most especially the availability of water. So we look for a

FIG. 3.1
A Pala'wan woman holding cassava cake, an unappetizing but consistent staple impacted by only the most severe drought conditions

spring, and then we built a well from that spring so we can just have a source of drinking water."

Lack of drinking water was only one direct impact on life in the uplands. One *panglima* noted the expansive forest fires that occurred with regularity during the drought: "When there was extreme heat, it caused a wildfire starting from Bangkalaan to Rio Tuba. The water was all dried up, and even the Marangas River was affected by fire . . .[3] and all the banana plantations were burned." While wildfires presented an immediate threat to Pala'wan homes and livelihoods, such dangers were further compounded by the effort of the municipal government to suppress further fire outbreaks. On Palawan and elsewhere in the Philippines, foresters routinely blame indigenous people and their agricultural practices for the outbreak of wildfires. Rather than emphasizing the environmental impact of fires or the pressure on indigenous Pala'wan lives and livelihoods, the municipal governments of Brooke's Point and Bataraza saw wildfires at the time as a threat to lowland paddy fields and cash crop plantations that bordered the foothills of the mountain range. Local government officials throughout southern Palawan believed the swidden practices of the Pala'wan were directly culpable for the conflagrations in the uplands that could potentially spread to the lowlands, in addition to damaging the watershed from which paddy rice irrigation systems and migrant households drew their water supplies.[4]

As the burning of fields was impossible to conceal, Pala'wan had difficulty in clearing fields for the following season, further impacting productivity in swidden plots. As recounted by an elder *panglima*, local police held upland Pala'wan households collectively responsible:

> During the El Niño there was a forest fire that happened here, and then they called for a meeting in the elementary school, and then a police offer was also there. His name was Chris Po from Brooke's Point, and then he asked, "Where is that mountain, and that place?"
>
> So I told him the name of the mountain and the place, and he told us that after two days if the fire did not subside, he will force the people living in the mountain to transfer here to the lowland.
>
> And then I answered him, "If you will force them to live here, where will they get their food? And they don't know other livelihood, and they don't have carabao to plow the land. In short, they will have no food for their family."

And he answered back, "Just be careful, because you are really fond of burning the forest."

I answered back, "How can we plant there if we will not burn it?"

He told us to use rake, and then I said it's not possible to do that because the trees are big, the same size as the thigh. And then he just told us to clear the boundary thoroughly before burning,[5] and if the fire goes beyond the borders [of the plot], it is our responsibility to stop it. But some of the people here are hardheaded, so I told them if they will not follow, the police will get them, so they followed afterwards, and then after two days the forest fire stops. That's why today very seldom that people here are burning the forest, so we can protect it from destruction.

Swidden agriculture was not the only livelihood pursuit negatively impacted by drought. By 1997, most Pala'wan households had come to rely on a suite of market-based activities to facilitate household reproduction. This meant wage labor, cash cropping, and to a lesser extent, emerging efforts to move into paddy farming. Severe drought had a significant and devastating impact on all these activities. Wage labor in lowland paddy fields, an increasingly key source of cash income for many Pala'wan households since the 1960s, was particularly impacted as the Inogbong River ran dry and starved lowland paddy fields of irrigable water. Though some lowland migrant farmers were able to continue cropping schedules by exploiting small springs, the majority of paddy rice farming in the lowlands, and the supply of waged employment, ceased. As one Pala'wan man related, "No, because the owner of the *basakan* is also affected, there is no water in their *basakan*; that's why they don't hire people." A Pala'wan woman highlighted the doubled impact of reduced cash income and relatively high rice prices, explaining, "There is still some work but very seldom, and if they [lowland farmers] need someone to work, they want to finish the work in one day only.... And also all of Bataraza is affected during that time, so there is a shortage of rice, and we imported rice from the neighboring countries and it cost 120 pesos per *ganta*."

Other sources of cash income or subsistence that could potentially buffer against swidden failure, including emerging engagement with cash crop markets and indigenous experiments with paddy rice, were likewise diminished. A Pala'wan man described the broad impact of the drought: "We really experienced a hard time. Our banana did not bear fruits. The rice was burned completely, except for the paddy fields that have enough water. But in *kaingin*, no one here had a harvest during that season. And all the people here were so very hungry. We have cassava if we planted it near the river banks, and then the relief goods from the government come and was of some help to us." The extended drought in Inogbong therefore had a significant effect on Pala'wan productive capabilities that was felt across all livelihood practices. As swidden plots, upland and lowland paddy fields, and other income-generating activities successively failed over the course of 1997 and 1998, Pala'wan households experienced dramatic inability to secure basic subsistence from their own agricultural practices or cash income.

PALA'WAN RESPONSES TO FOOD SHORTAGE

As stocks of existing food sources were exhausted or failed and the 1998 cropping season was delayed beyond March–April of that year, Pala'wan households recalled obtaining sufficient food through a diversity of sources.[6] While limited relief goods (such as canned foods) were distributed to Pala'wan households through municipal and *barangay* calamity funds, the most important source of subsistence was the collection of wild food sources. Primarily, this meant the gathering of tubers, fruit crops, sago, and to a lesser extent, wild grains and vegetables. One Pala'wan man recalled, "I don't know why it happened, but all I can remember is that all the people were hungry because no food was available. Even the banana and coconut did not bear fruit. And my children only ate mango for breakfast, lunch, and dinner.... No other food was available, sometimes the *ubod ng saging* [heart of the banana tree]." Another man further explained, "And then our food is the root crop, mostly *kedut* [*Dioscorea hispida*, a small, white-fleshed tuber]. We soak it in water for three days before we can cook it. The *sarawak* [a wild tuber] is a big help for us during that time. And another root crop is the *apari*, but also the *batbat* [sago or other edible starchy palm]. And another one is the *ilus*, and it is a black and long kind of root crop."

Many of these food sources, while easy to access from nearby forested areas, require extensive processing to remove poisons and render them edible. For example, oral histories point to *Dioscorea hispida* as a common source of food as the drought progressed, which, though plentiful, must be grated or sliced and soaked in a flowing river for several days to remove excessive alkaloids,[7] a process that became difficult as the Inogbong River ran dry. Other sources of wild food, such as wild fruit, required traveling great distances to gather in any effective numbers, as one *panglima* recalled:

> We get *badak* [*Artocarpus champeden*], and other fruits....
> We leave here six in the morning, and then we get home seven in the evening ... girls and boys, young and old, as long as they can walk, they are helping to gather the *badak*. The young ones can only carry seven at a time because it is the size of a thigh. We collected the ripe fruit, and we ate that right away, and we cook the seed [of the fruit] and smash it, then we ate that too. [If it's not ripe,] we boiled it and mix it with coconut milk; if the coconut milk is not available, we just boil it and put some salt. And that will serve as our breakfast and dinner every day.

Oral histories of the time suggest that such wild foods formed the bulk of subsistence during the drought. Like domesticated root crops and other nonrice cereals, wild foods were readily available but extremely undesirable as everyday staples. Wild tubers, aside from the frequent need for processing, are generally thought to have an unpleasant taste when compared to their domesticated counterparts.[8] Indeed, the description of 1997–98 as a time of hunger refers not only to the difficulty in obtaining sufficient subsistence but also to an extended period without rice. Many Pala'wan thus went to considerable lengths to facilitate the consumption of imported rice, and the failure of both upland and lowland agricultural production was compounded by high prices and poor-quality National Food Authority relief rice.[9] Households that had more successfully integrated into lowland markets found their cash-generating livelihood activities—waged labor, paddy farming, and cash cropping—most compromised by lack of rain and this meant that alternate sources of cash income were needed if purchasing rice was to continue.

The forest served as a source not only of emergency subsistence foods but also of potential income generation. The extraction of forest products for sale remained at least partially viable during 1997–98 for many Pala'wan people. For example, facilitating small-scale illegal logging, long a source of cash income for indigenous peoples throughout Palawan Island, was an important buffer against food shortage for some households. Though the rise of somewhat effective state enforcement of timber concessions in the 1970s and '80s had tempered the trade to a degree, oral histories suggest that increasing desperation among both indigenous people and migrants prompted a mild resurgence in illegal logging in the area. Rather than gather and sell timber themselves, some Pala'wan men acted as guides for migrant loggers by locating and assisting the transportation of raw timber.[10] One indigenous man recalled purchasing "smelly" National Food Authority relief rice in the lowlands with income derived from logging: "We really experienced hunger during that time; it is not possible to get food here.... We could not continue to live if there were no people using chainsaws; that means we were loading the wooden boards. And the smelly rice cost eighty pesos [per *ganta*] at the time, so we bought it with the help from moving the wooden boards."

The collection of *bagtik* also proved to be an important source of income for some households living in the forest interior, whose members had sustained experience collecting forest products over great distances. In an ironic turn, the usually strenuous collection of *bagtik* became somewhat easier, as dry weather allows for the effective transformation of heavy loads of resin that are otherwise impossible during monsoonal rains. As one Pala'wan man living in the forest interior, still relying on the sale of *bagtik* for cash income, recalled, "[During that time] we sold *bagtik*, and then we bought rice in the market ... and until now we are still collecting *bagtik*. Before to Rodriguez, but now to Panglima Burit."

Though gathering wild fruits and tubers and purchasing relief rice were vital means of obtaining food, indigenous families initially secured subsistence from upland swidden plots. While reports of total crop failure were widespread, drought conditions impacted households and swiddens located in the forest interior to a lesser degree than those closer to the lowlands, as upland topography is palpably moister, cooler, and less likely to dry completely during drought. In these areas, small crops of existing cassava and other roots survived the initial lack of rain in the area.[11]

For those families in the foothills, leveraging social relations and norms of reciprocity with their less impacted kin in the forest interior was a valued source of subsistence. This was primarily focused on root crop exchange.[12] One Pala'wan woman considered drawing food from upland kin as part of a range of strategies to ensure adequate food: "I didn't have piggery that time, so we really experience hunger. But my cousin living in the mountains, they still had cassava, so we go there and ask for cassava. And not only that, if there is someone who will hire laborers, I will also work.... Sometimes we also get *sarawak*, and then *apari* [varieties of wild yam], and it has many thorns, so we must be careful when getting it." In a similar vein, a man recalled that "some of us were able to plant some root crops so they have food. But for us, we just ask some food to our relative and neighbor who still have food."

The moral economy that exists between Pala'wan residing in the forest interior and those in the foothills became important in not only providing an additional source of food in times of hardship but also ensuring the long-term survival of swidden production after the drought had ended. Beyond securing basic subsistence, many households leveraged the same social relations to secure seed rice for the following year's swidden plot. As most households had experienced a total failure in rice crops, no seed was available for the following season. In other households, despite extreme hunger, many Pala'wan preserved small stocks of seed rice in anticipation of the following year's cropping. Households in the forest interior kept aside what little could be harvested as seed stock rather than eaten. Some indigenous people recalled hand-watering small rice fields to secure future supply, rather than consumption. It was this conservation of seed stock that facilitated the following year's harvest for many households in the *barangay*, who leveraged the same kin relations in the forest interior to secure rice seed. One Pala'wan man explained:

> There are *kaingin* that were affected, and there are also some that were not affected. So, we get our seedlings from our neighbor and a nephew who did not suffer as much. Some of us were able to harvest [and have seed stock], and some others just bought their seed from the other farmers. In that time, there is a paddy field where we buy the seedling and it cost one hundred pesos per *ganta*. In *sumbang* [incest] cases, the curse only impacts the place where it happened, so that's why in some other places they were able plant seeds.

In other cases, however, farmers traveled long distances simply to reacquire rice seed stocks. Many households reported visiting relatives in distant *barangays* of the municipality, while still others traveled to the opposite side of the mountain range to, for example, Culasiun in the adjacent municipality of Rizal. As one woman recounted, "I have relatives in Tigwayan [an area on the opposite side of the mountain range], and they were not as affected by the drought, so they were able to save some of the seed. I asked them to give me even a little amount of seed."

As such, while the impact of the 1997–98 El Niño–induced drought on agricultural production was severe, Pala'wan households had a wide range of strategies to obtain subsistence. The extensive use of wild root crops and varied income-generating activities meant that starvation from hunger was limited, at least to the extent that such accounts do not feature prominently in oral history. This seems to have been the case throughout the island. Media reports focused on the drought attributed any fatalities to exposure to forest fires rather than hunger (Florece et al., 2002). By drawing primarily on forest resources, households were able secure an adequate amount of food, though enduring a diet that was unpleasant and devoid of socially and culturally valued rice. The careful maintenance of seed stock and its subsequent dispersal facilitated the reestablishment of swidden rice in 1998. Alongside the ingenuity of indigenous households in dealing with extreme weather conditions, these oral histories also speak to the role of diverging indigenous livelihood practices in mediating how the drought was experienced and the immediate coping strategies available to different households. Though intense heat badly impacted swidden fields throughout the *barangay*, some Pala'wan families were able to draw on stocks of root crops as a source of emergency goods. Households who had enthusiastically grasped new livelihood activities found their productive capabilities sharply curtailed by the drought with few alternatives for food procurement.

Pala'wan experiences with drought suggest that trajectories of livelihood change in which indigenous families increasingly abandon swidden agriculture may lead to acute vulnerability to drought among many such households. In oral histories, wage labor in lowland paddy fields was extremely rare during the 1997–98 drought as irrigable water available for migrant paddy fields became scarce, choking an important source of cash income for Pala'wan families.[13] Indigenous households that specialized in paddy rice and cash-cropping farming found themselves similarly vulnerable to El Niño conditions as the drought severely impacted these forms of agricultural production.[14] The impacts of the 1997–98 drought highlight configurations of livelihood and residence that are particularly fragile and yet have all become central to household reproduction not only in terms of providing cash income but in many cases also as the primary sources of subsistence.

MEMORIES OF HUNGER

Traumatic experiences of hunger and struggle could be expected to become the basis of strategic household responses that shape livelihood activities. To some extent, contemporary Pala'wan livelihoods are reflections of these past experiences with extreme drought. This is particularly true of households headed by women and men who had already established their livelihoods at the time of the 1997–98 El Niño. Though El Niño events regularly mediate the length and intensity of monsoon activity, most indigenous people considered the 1997–98 drought exceptional in its impact, and older Pala'wan suggest that a drought of such intensity was unprecedented in the area, at least since their earliest recollections, beginning in 1940s and 1950s.

However, the Philippines and the municipality of Bataraza periodically experience El Niño–induced droughts of similar intensity to that of 1997–98, in line with shifts in the Southern Oscillation Index.

Because of the severity of the drought, the events of 1997–98 have become a pivotal point in indigenous narratives of environmental change and livelihood transformation. Indigenous families often insist that the impacts of the drought were unprecedented and frame livelihood decision making in relation to climatic patterns. One *panglima* articulated a common perspective, that, "our parents never experienced that extreme heat before; what they experience is the normal dry season. For example, after they burn [their plot] today, by tomorrow it will rain. But right now, it's very different."

As most Pala'wan attribute the unique extremity of the 1997–98 drought to incestuous behavior or other kinds of linear moral decline rather than cyclical weather patterns, the widespread perception of declining moral restraint within the community means that many indigenous people understand El Niño conditions to occur at increasing intervals. Because of these perceptions of environmental decline, many households anticipate future El Niño events roughly in accordance with the projections of climate scientists around the world and have adjusted their livelihoods accordingly. However, though Pala'wan people widely recognize the possibility of El Niño events, they typically limit efforts to minimize the impacts of future droughts in household subsistence to shifting cropping patterns within swidden plots. In response to the intense drought, many Pala'wan suggest that they will "just plant" (*megwulak lang*), or as expressed in Tagalog, "really planted [a lot]" (*magtanim talaga*). While both the Tagalog and Pala'wan verbs *magtanim* and *megwulak*, respectively, may refer to any crop planted within a swidden field, what is more specifically meant is that households have a density of crops known to be drought resistant rather than increasing the size of the fields or increasing the overall density of crop production.

As cassava is one of the few crops to survive long periods of drought, especially in the moister forest interior, informants suggest that they strategically increased cassava production in the years following 1998. In a discussion with one Pala'wan man about cropping decisions, he justified emphasis on root crop production by noting that such events may recur: "We also heard that it [the El Niño drought] might happen again in the coming years, so what we are doing right now is we are planting more root crops, coconuts, and other fruit trees so that if that time will come, we have something to eat." One *panglima* in the forest interior noted that the rise of root crops within swiddens is a response to the hardship experienced collecting wild fruits from the forest during the previous drought: "From then, when the El Niño happened, to now we planted a lot so we don't need to get *badak* [from the forests] again. The reason is that we planted banana and cassava."

However, this strategy was not confined to the forest interior of the *barangay*, where swidden production is more important to household livelihoods. In the foothill hamlet of Papan, a Pala'wan man noted that the emphasis on cassava and root crop production has become institutionalized within communal cropping schedules:

> We conducted a meeting, and I suggested that we take advantage of the good weather we have right now to plant in our *kaingin* the crops that can survive for a very long time,[15] so that if it will happen again, we can dig it up and we have food to eat . . . cassava, banana, sweet potato. I was living in Ilay at that time. I have half a hectare of field there. I was working in Brooke's Point that time, and when I came back, it was being cultivated by the lowlanders, so I lost my *kaingin* then. . . . Before, there were only a few crops in that field, but now we plant thousands.

Self-reported crop data from household surveys reveal a significant emphasis on cassava production among Pala'wan households in both the foothills and the forest interior, which routinely aim to cultivate over a thousand individual plants each year. While there are no comparable data from before 1997, many informants suggested that this amount is the result of a transformation in the emphasis given to root crops with swiddens, often explicitly in response to extreme drought. For example, when asked if such large numbers of cassava were common during 1997–98, one Pala'wan man living in the forest interior highlighted experiences of hunger in shaping these kinds of decisions: "No, that's why before we experience hunger, for example [my father], he only planted 500 pieces of cassava for one year, so I said in my mind, it is not enough for the whole year, so I've decided to plant 10,000 pieces." He continued, reflecting on experiences with subsequent, smaller periods of drought:

> The same with my father-in-law, we did not have a hard time during that [subsequent] drought because we have planted 10,000 cassava plants, and even the people from Papan, we were able to help them. . . . But even then, we were not able to consume it all. There is one person from Puerto [Princesa] who

came here just to get some cassava. The reason why we have planted so many [is] for us not to be hungry if ever the El Niño will come again ... until now, that is why we have so much cassava available.

He notes that indigenous households in the foothills, and even buyers from as far as the provincial capital, came to rely on his field. Another Pala'wan man emphasized that the performance of crops during drought events in the years following 1998 validated the efficacy of these strategies: "It happened again, but it's not difficult compared to the first one because we already have planted some root crops."

These conversations with Pala'wan people suggest that swidden, and specifically an emphasis on planting large numbers of drought-resistant crops, sits at the center of contemporary Pala'wan efforts to buffer subsistence against severe drought. Though crops such coconut and banana did fail during the 1997–98 El Niño, Pala'wan farmers see them as hardy and drought resistant and place them alongside cassava and other root crops as generally safe but unappetizing staples. But while Pala'wan farmers explicitly identify drought as a motivating concern in swidden cropping patterns, broader livelihood aspirations are not built around anticipation of future extreme climatic events. Swidden is being abandoned entirely by a small but growing number of households, whose members favor specializing in paddy farming, wage laboring, or other activities that can secure cash income. For others, swidden is, while still practiced, an increasingly peripheral element within broader familial aspirations for regular supplies of rice and household goods.

The precariousness of dependence on nonswidden activities is not unrecognized. For example, one *panglima* in a hamlet whose families are heavily dependent on wage laboring articulated concerns over the vulnerability of a livelihood based on wage labor. When asked how households who have abandoned swidden agriculture and now rely on purchasing food rather than producing it themselves would cope, he highlighted the collective vulnerability in such patterns of livelihood: "It's up to them what they will do. They can buy food if they have money, but here we don't tolerate stealing, so if someone needs food, we just give them food so that they will not steal." This man and his wife had abandoned swidden production entirely some years earlier and now specialize in paddy rice farming. They sought to buffer their vulnerability to failures in paddy rice by negotiating a share of cassava production in the swidden fields of upland relatives, which may be tapped in times of need. These attitudes expose the continued perception among many households that swidden is an important subsistence buffer against disaster. Such practices are, however, the exception to the rule. When considered within broader trajectories of change since 1998, many Pala'wan households clearly hold other aspirations higher than mitigating future drought risk.

CONCLUSION

Oral histories from 1997–98 in Inogbong reveal the range of resources and strategies leveraged in times of crop failure to ensure basic subsistence, centered primarily on accessing surviving root crops from swiddens in the forest interior, foraging for wild fruits and tubers, and purchasing rice with cash income gained from limited wage labor and the sale of nontimber forest products. Careful maintenance of fields and efforts to reacquire upland rice stocks allowed households to reestablish their swiddens in the following year, and over longer timescales, households have transformed their cropping patterns to now heavily emphasize drought-resistant varieties such as cassava and banana.

Pala'wan experiences during 1997–98 suggest a broad capacity of indigenous households to cope with food shortage amid extreme drought, as well as reestablish and modify swidden practices following near-total crop failure. However, since this time, Pala'wan livelihood aspirations have increasingly abandoned or marginalized these kinds of practices. Concerns regarding extreme drought have been largely subsumed by more pressing issues, such as decisions regarding livelihood practices are subject to shifting, competing, and often contradictory demands. Pala'wan households are willing to shape their swidden practices in response to the risk of future drought events, planting large quantities of root crops as a buffer against variance in rainfall. However, in the context of local perceptions of swidden decline and ongoing forestry efforts to reconfigure swidden agriculture, many Pala'wan continue to move into activities that are extremely vulnerable to extended drought, placing higher value on issues such as self-sufficiency in rice and availability of cash income over resistance to extreme weather events in the future. Wage labor, paddy rice production, and cash cropping have all now become vital aspects of indigenous livelihoods since 1997–98, particularly in the foothills of the *barangay*. In addition to livelihood change, the moral economy of food exchange across the upland-lowland divide, which sustained many households during the drought, is under threat as households have come to concentrate their residency in the foothills, closer to wage-laboring opportunities. For many Pala'wan, daily imperatives to seek a future outside of swidden agriculture and residence in the forested interior overshadow the potential impact of a similarly intensive El Niño event.

How can these kinds of examples fit within critical narratives of vulnerability that are concerned with identifying not only causality but also politically loaded culpability (Ribot, 2014)? One response, common throughout much normative literature focused on analyzing the impacts of climate change, has been to blame rural peoples for their own fragility. Within literature concerned with the vulnerability of rural households, these kinds of decisions are sometimes framed as "maladaptive." A maladaptive framing places blame on farmers themselves, emphasizing the shortsightedness of household members and, consequently, the need for outside intervention. The prescription of government intervention should be treated with caution, as disaster recovery efforts have been the means through which the state has inserted itself further into the lives of rural communities (Solway, 1994). Not only are these kinds of explanations morally unsettling, especially if applied to indigenous peoples, but they are also often empirically untenable. Pala'wan men and women have clear understandings of the cyclical nature of El Niño droughts that have communicated through familial networks and hamlet meetings the enduring nature of this risk. Outsiders, who develop and design far-reaching programs of livelihood on the basis of staggeringly flimsy knowledge about how people actually live their lives, have little to offer.

A range of political economic trajectories and governance dynamics has partly shaped the conditions for a recurring subsistence crisis. Processes of frontier settlement have marginalized indigenous peoples throughout Palawan and produced troubling relations of dependency with migrant lowlanders. Undoubtedly, the punitive pressure exerted by forestry projects over time has also influenced the decision to move away from or de-emphasize swidden agriculture. At the same time, Pala'wan experiences of livelihood change also defy efforts to abstract blame solely to impersonal state forces, such as forest management efforts. The reasons for livelihood change in Inogbong and on Palawan Island more broadly are far more convoluted, and indigenous peoples themselves actively take part in the uneven decline of swidden agriculture, or "customary" life. Considering new household needs for cash income and in the context of declining productivity, indigenous peoples in Inogbong have increasingly adopted a diverse range of practices as a means of securing a good life. Despite the concurrent value of swidden as way of buffering weather-related risk, there is a greater desire for household goods and ways of living that exist outside of mobile and forest-reliant modes of livelihood. Yet such desires and aspirations for market-based livelihoods are hardly entirely organic or politically neutral. These kinds of trajectories are also the result of shifting environmental subjectivities and moral reevaluations of upland forest living that arise both from *and* beyond forest governance projects. Many Pala'wan households willingly espouse the rhetoric of the PTFPP to justify transitions away from customary agriculture, and they discipline themselves and their neighbors to sedentarize and commercialize their livelihood activities away from the forested uplands.

These complexities have implications for how scholars conceptualize vulnerability, and just what vulnerability entails as a moral project. Rather than reproducing reductive narratives of blame that point to either the deficiencies of rural households or abstract historical forces, positioning long-term Pala'wan responses to drought in the context of struggles to make and change livelihoods can attend to the ways in which vulnerability may be viewed as insidious. Doing so does not necessarily provide a kind of clear moral resolution, but it can point the way forward to a more nuanced analytic of vulnerability that can accommodate forms of diffused and distanced culpability relationally produced among a range of actors.

CHAPTER FOUR

El Niño and Incest

I ENTERED into my research in Inogbong determined to find a discrete body of environmental knowledge that was consistently put into practice to address changing weather conditions. I initially sought out knowledgeable Pala'wan people who could satisfy my intellectual desire for expertise that would provide evidence of a key thesis embedded in much global environmental governance: indigenous peoples, while victims of climate change, could also be the saviors of environmentally disconnected Western societies. I was already well versed in the critical literature on indigenous knowledge that emerged from the 1990s, which had critiqued notions of stable and uniform bodies of "knowledge" that could be neatly captured and packaged through academic research. Yet heady fantasies of climate change knowledge edged out these critiques and produced a wish for the stereotyped and romantic figure of indigenous person as climate expert.

One my early interviews with a Pala'wan man revealed how difficult finding this kind of figure would be. After a series of questions surrounding livelihood practices, with a topic I already had some familiarity with from conducting research with indigenous people in central Palawan, I asked, "Nagbago ba ang panahon? Ang init, ang ulan?" (Has the weather changed? The heat, the rain?) He looked at me quizzically, replying, "Well, of course, the weather changes every day." Learning how, and whom, to ask the right questions about the weather took some time. Slowly attending to the complexities of environmental knowledge as referenced to specific livelihood activities helped elicit aspects of the expertise I had originally been so fixated on finding. However, what emerged from this process was far from a validation of my expectations and was, instead, deeply morally disorientating. For Pala'wan people in Inogbong during my fieldwork in 2011 and 2012, the onset of the southwesterly *barat* monsoon that arrived every March or early April to nourish newly planted rice and vegetables had been predictable. By this time, Pala'wan people of all ages consistently suggested that they could no longer accurately determine when—whether reckoned by calendar month or astronomical signals—they should burn their fields and plant their crops in anticipation of rain. When considering changing weather in interviews and casual conversation, indigenous people placed blame squarely on their own increasingly immoral behavior. They frequently saw this as part of a broad sense of social decline. Children no longer obeyed their parents, neighbors no longer shared with each other as they once did, the earth itself was becoming "old" and could no longer ensure regular cycles of rain and sunshine. The most dangerous manifestation of this trend was the widespread perception that incestuous relationships among Pala'wan people in Inogbong had become common, leading to divine disfavor, cosmological imbalance, and ultimately, new and dangerous weather conditions.

The calculation of victimhood through specific conceptualizations of vulnerability is central to climate change narratives. Vulnerability, however, is only half of this larger and often contradictory imaginary that positions indigenous peoples in relation to global environmental change. As anthropologists Lucas Bessire and David Bond (2014, p. 450) have suggested, indigeneity offers a "site for the philosophical and moral redemption" of the contemporary Western world. Nowhere is this perhaps more obvious than in climate change discourse, where indigenous peoples often offer the possibility of a pragmatic redemption for Western societies in the form of climate change "mitigation" or "adaptation" strategies. In a recent interview, Victoria Tauli-Corpuz, the UN special *rapporteur* on the rights of indigenous peoples and an indigenous woman from the Philippine Cordillera, articulates the basis of this redemptive vision: "Since indigenous peoples are contributing significantly to climate change mitigation by living simply, further developing and practicing their traditional knowledge systems and by conserving and sustainably using their forests, agricultural lands, coastal resources, among others, their rights to continue doing these mitigation measures should be respected and protected" (Internews, 2018).

Strategic essentialisms of indigenous peoples as ecologically harmonious in climate change discourse are often functionally operationalized through the promotion of "indigenous knowledge" and its many variants. At the same time, this redemptive hope is dependent on an extremely narrow vision of what indigenous knowledge should look like and entail. In situations where indigenous peoples' understandings of the weather do not align with scientific climate data, or they blame themselves for changing climatic conditions, these views are often cast as barriers to the pragmatic challenges of dealing with climate change impacts or simply ignored.[1] They are dumped in an "epistemological ghetto" (Brosius, 2006), a conceptual space reserved for knowledge claims that do not fit neatly within the utilitarian policy aspirations of states and global policy networks, or that defy the simplified expectations of indigenous peoples as expert and harmonious stewards of nature.

How else could these understandings of climate change be interpreted? What other possibilities might they offer for scholars grappling with issues of indigeneity and environmental change? Anthropologist Virginia Nazarea (1999, p. 5) has suggested that it is no longer possible for anthropologists to naively ignore the politics of knowledge that operate among informants and often result in fluid and conflicting accounts of "traditional ecological knowledge" within a seemingly homogenous community. In the context of a more recent "ontological turn," "taking seriously" forms of alterity might mean carefully focusing on the everyday politics and positionality through which knowledge claims are deployed and authorized, rather than reifying bounded indigenous cosmologies (Astuti, 2017, p. 106). Turning away from an apolitical, static, and singular vision of indigenous knowledge, and of indigenous ontologies as monolithic, could instead mean attending to the ways that wider struggles over place intersect with and shape the ways that people think about weather.

Pala'wan discourses of climatic change, though rooted in the language of ritual and tradition, cannot be viewed in isolation from the Philippine state's efforts to reshape the social and environmental contours of forest landscapes in Inogbong. How and why Pala'wan talk about climate and climatic change is imbricated with ongoing debates among indigenous peoples over how their behavior and environmental resources should be managed and what role the Philippine state should play in these processes. A focus on swidden livelihoods—as a sensorial human-weather interface and the contested object of governance—speaks to the embodied politics that surrounds climate change for Pala'wan households.

Pala'wan households construct an idealized pattern of weather in relation to the annual swidden cycle and negotiate climatic uncertainty through seasonal markers. By understanding and experiencing weather patterns primarily in relation to livelihood, Pala'wan households hold consistent expectations of certainty or normality in weather patterns. In the context of a cosmology that links Pala'wan social behavior with environmental functions, the climate discourse of indigenous peoples in Inogbong takes on a moral character as social dysfunction within the community is manifest in irregular and extreme climatic conditions. The way indigenous people understood deviations from these climatic norms in explicitly moral terms during 2011–12 involved a perceived link between rising incest (and other moral degeneration) and worsening climate. These perceptions of increasing incest and changing climate occur, however, within histories of state forest management and perceptions of emerging social differentiation among indigenous households. Though articulated through language that connects weather to ritual and tradition, Pala'wan strategically and fluidly reorganize these understandings of climate in line with contestations over political authority and local socioeconomic inequities that have arisen in tandem with forest management efforts in the *barangay*.

THE SOCIAL CONSTRUCTION OF CLIMATE

In meteorological science, the climate concept refers to the statistical representation of temperature, precipitation, and wind speed over long time periods. These models are produced by averaging variables over, typically, thirty-year periods to generate a representation of how the weather might unfold on any given day of the year. If strictly understood as an assemblage of these technoscientific techniques, meteorological notions of climate "travel" poorly outside of scientific circles. Geographer Mike Hulme (2017, p. 6) offers a more expansive notion of climate as "the human sense of climate that establishes certain expectations about the atmosphere's performance and how we respond to it. The idea of climate cultivates the possibility of a stable psychological life and of meaningful human action in the world. Put simply, the idea of climate allows humans to live culturally with their weather." The climate is therefore a "normalizing idea" related to the management of expectations and the stabilization of uncertainties rather than a purely objective representation of weather patterns and processes. This definition suggests that human societies culturally construct climate, as a set of ideas of normality surrounding the weather, in relation to a range of anxieties at the interface of weather and human endeavors. Studies of climate as a cross-cultural phenomenon, however, tend to reduce weather knowledge to a function of an often-essentialized and static set of livelihood practices. Through this lens, the observations of indigenous peoples are produced in narrow reference to specific kinds of activities: farming, fishing, hunting, and so on. However, notions of livelihood can refer not only to a routine and sensorial set of practices, but also to the politics of how and for whose benefit people should live in tropical forests. For Pala'wan people, therefore, anxieties of weather are rooted in not only the cycles and specificities of swidden in Southeast Asia's tropical forests but also the politics that envelop and aim to transform these practices.

While nonswidden climatic markers are sometimes referenced (e.g., the flowering of certain kinds of trees, mushroom growth), Pala'wan people in Inogbong largely articulate weather in relation to specific swidden practices. Their concerns regarding weather are primarily centered on the timing of clearing and burning of vegetation within a plot and the subsequent planting of rice. The precise timing of the burning practices requires fairly accurate knowledge of when rains will occur; ideally, a plot is burned one week prior to planting rice, and rice is planted one week prior to the arrival of the monsoonal rains.[2] Pala'wan households navigate the timing of burning and planting practices through the construction of seasonality.

Perhaps unsurprisingly, seasonal divisions and markers are oriented around swidden production. Pala'wan seasonality divides the year into wet (*barat*) and dry (*bulag*) periods. The *bulag*, beginning in January, brings hot and dry northeasterly trade winds. Sometime in April, the *barat* period begins and marks the arrival of the southwestern monsoon that is responsible for much of the island's rainfall. Pala'wan divide the *barat* period into five stages of increasing intensity that roughly mirror statistical rainfall averages: (1) Wind of the Duduy (chick), lasting from May to June; (2) Wind of the Indu' (mother hen), lasting from July to August; (3) Wind of the Buntal (globefish), lasting from September to October; (4) Wind of the Nyug (coconut), occurring during November; and (5) Wind of the Pungur (leafless tree), during December.[3] The precise alignment of *barat* and *bulag* periods to calendar months is subject to variation depending on an individual's familiarity with celestial patterns and Western calendar months; what is presented here is the most common idealization of climatic periods.

These seasonal changes and their associated swidden activities correspond to a range of celestial phenomena that guide agricultural practices. For example, the beginning of the *barat* period is marked by the heliacal fall and rise of the constellation Indu' (part of the Western constellation Auriga that resembles a hen). Increasingly the Western calendar has become more prevalent in the interface between livelihood and climate. Whether Pala'wan frame climatic patterns in terms of Western calendar months, indigenous celestial observations, or both, the expectation of seasonal normality remains consistent. That is, the dry season is expected to be dry, and the wet season is assumed to increase steadily in intensity before the arrival of the following dry season. It is this threshold between *bulag* (dry) and *barat* (wet) periods that occupies the central focus of swidden-practicing households, as critical elements of shifting cultivation (clearing, burning, and planting) are dependent on the precise timing of the transition to the rainy season.

However, both statistical monthly rainfall averages and Pala'wan seasonal divisions present a misleading pattern of annual regularity. Climatic data collected close to Inogbong from a Philippine Atmospheric, Geophysical and Astronomical Services Administration (PAGASA) weather station from the 1950s to the 2000s reveal a much more erratic pattern of rainfall. If averaged over forty-four years, rainfall data from March would suggest a consistently low rainfall (less than 50 mm). However, if March[4] is disaggregated and examined over the same period, it is clear that there is intense irregularity in rainfall on a year-to-year basis: in some years March may experience relatively high rainfall (well over 100 mm every few decades), and in others it may experience effectively no rain. What is consistent is the inconsistency. Meteorological research focused specifically on the Philippines has emphasized that variation in the Southern Oscillation Index is linked to these significant departures from expected seasonal rainfall averages in addition to the early or late arrival of the southwest monsoon. Considering the SOI relative to March rainfall over time reveals a clear though erratic relationship between the Southern Oscillation Index and local weather patterns. While positive and negative SOI values are associated with more or less rainfall, respectively, the strength of this association varies from year to year, consistent with broader scientific understandings that connect ENSO "events" to the intensification of seasonal averages (Allan, 2000, pp. 4–5). While El Niño events frequently produce an elongated dry season, La Niña events are linked to the early arrival and greater intensity of monsoonal rains.

While climatic data suggest that monsoonal onset and intensity are extremely variable, Pala'wan people generally possess an expectation of regularity in climatic conditions, framed in terms of optimal swidden productivity. Pala'wan often assume that weather patterns will and should align with seasonal divisions and monthly expectations of rainfall. These climatic norms, which roughly mirror scientifically collected data and are expressed in terms of relative amounts of rainfall, are understood not only to reflect environmental processes but also to depend on the "good" or "bad" properties of social relations, which ultimately regulate climatic conditions.[5] Pala'wan cosmology points to an intimately socialized nature in which humans and nonhumans regularly interact in reciprocal and carefully ritualized relationships. As a result, Pala'wan people understand deviations from ideal rainfall conditions as divine disfavor or curse resulting from improper or immoral human actions. The stability of ecological systems is therefore dependent less on abstracted cosmological principles than on interaction between humans or between humans and divine agents grounded in everyday social relations and agricultural practices. As Pala'wan frequently frame ecological functions in interpersonal terms, weather events take on a moral character—good behavior engenders good weather and, in turn, optimal swidden yields.

INCEST AND WEATHER

During my fieldwork and discussions with informants regarding swidden practices in 2011–12, a clear and consistent discourse regarding transformations in weather patterns emerged: whereas in the past, the onset and duration of the wet and dry seasons were more certain or acted upon in line with celestial or calendar markers, seasonal change has become increasingly unpredictable. Though this climatic variability can impact various components of the swidden cycle, as noted above, what concerned most farmers was the ability to burn their plots. It is difficult to burn a swidden plot without a reasonable expectation of when monsoonal rains will arrive at the threshold between *bulag* and *barat* seasonal divisions. As a result, it was common for farmers to suggest they have not been able to burn their plots in recent years and that this had significant implications for rice productivity.[6] One *panglima* articulated a common belief that inclement weather is the result of *sumbang*, or cases of incest:

> I really can't understand the weather conditions that we have right now. Sometimes there will be rain and sometimes extreme heat—it is really becoming unpredictable. Before, when there is no *sumbang* . . . it was easy to predict if it will be a sunny day or rainy day. But now, it is very unpredictable. Before, we looked at the stars in the sky as our guide for the weather; for example, before, when the month is April, we expect to have rain, but now, even it is June already, there no rain yet. I just asked Panglima Meyreg to try to settle the issue there [of *sumbang*]. . . . I don't know if it is settled, but I guess not, and all the *panglima* must be present when dealing with the *sumbang* case.

Throughout the uplands, customary leaders connected the occurrence of *sumbang* to an increasingly unpredictable climate (figure 4.1).[7] In many Pala'wan communities, Inogbong included, incest taboos prohibit sexual relations and marriage between close kin, both consanguineal and some

FIG. 4.1
Inogbong River during a flash flood. Pala'wan believe both intense rain and drought can be caused by incestuous relationships.

affinal. The precise boundaries vary, but incest taboos apply to close relatives up to and including first cousins, who are classificatory siblings, and prohibit relationships with in-laws across generations (Macdonald, 2007, p. 78).[8] For example, sexual relations between a man and his mother-in-law is a serious offense characterized as *sumbang*, as it violates both in-law and generational prohibitions. In contrast, my informants did not see sexual relations between a man and his wife's sister or brother's wife, while a serious act of marital infidelity, as an act of *sumbang*.

As *sumbang* is one of the most serious sexual and social transgressions, Pala'wan believe it causes extreme divine disfavor and, subsequently, various forms of environmental imbalance. Most frequently, Pala'wan understand *sumbang* to produce specific climatic events, trends, or conditions and associate the practice in local folklore with men of significant evil and physical prowess who engage in immoral behavior such as robbery and murder. Though informants also suggested alternate explanations of climate change, focusing on other forms of perceived moral or social dysfunction,[9] incest was the most clearly articulated reason for unfavorable climatic conditions, particularly among hamlet *panglimas* of the catchment, who consider themselves responsible for settling cases of incest in the area.

Why incest leads to a change in weather patterns is explainable in Pala'wan cosmology, which is grounded in a highly interpersonal and relational understanding of environmental processes. For many Pala'wan people in Inogbong, favorable environmental conditions must be continually maintained by ensuring good relations with a range of spiritual beings, deities, and ancestors through reciprocal and highly moralized exchanges. Incest is a significant social transgression that upsets the harmonious relationships between humans and powerful spiritual actors and can lead to environmental extremes, though the causal explanations that link incest to changing weather differ. Many *panglimas* suggested that these forms of sexual transgression displease the apical deity of the Pala'wan pantheon, Empu', who holds responsibility for environmental and cosmological balance.

Explanations vary from household to household: In some accounts, Empu' instructs the Tandayag, a giant serpent or dragon that inhabits the coastal oceans of Palawan, to travel into the mountains, bringing extreme heat or rain to the area where incest has been committed. Other explanations place greater emphasis on the Tandayag, who monitors Pala'wan sexual behavior. When incest is committed, the Tandayag is able to smell or sense the sins of the couple involved and travels under the earth to the area, causing intense heat.[10] Empu' may then send excessive rain to cool the area. Common to both explanations of the consequences *sumbang* is that unless the crimes of the incestuous couple are resolved, negative climatic conditions will befall an entire catchment rather than punishing solely the perpetrators. As a *panglima* in the forest interior explained:

> Yes, that is true, because the Tandayag lives under the soil, and he also feels the extreme heat, and that is also the cause of *sumbang* punishment.... If the Tandayag gets mad, I tell you that the whole Philippines will be broken into pieces, and it will sink into the water, and because the Tandayag is the creature God [Diyos] placed under the soil, and if there are earthquakes, it means the Tandayag moves. The story is like this: the Tandayag talks to God and asks for rain because he also feels the heat under the soil [from *sumbang*], so he ask for rain to cool him down, but the Tandayag is under the soil, so it will take time for the water to reach him, but on the surface there is flood already.

According to widely circulating oral histories, the crime of *sumbang* (depending on its severity) was until recently punished through ritual executions. *Panglimas*, both elders and the relatively young, describe these executions as elaborate rituals that reflected the changes in weather that were occurring and the specific depth of the incestuous crime. If, for example, there was an abundance of rain, the offenders were beheaded and the bodies left to dry. In contrast, if there was an excessive drought, the beheaded corpses were soaked in a river to bloat. After the ritual drying or bloating, the corpses would then be thrown into the ocean. In other accounts, both offenders were simply tied into a large basket with rocks and cast into the sea. According to older Pala'wan, at some point during the 1950s or '60s, the municipal government forbade the execution of incestuous couples. Since that time, they suggest, incest has been punished in less lethal fashions that fulfilled an equivalent ritual logic without violating state law. Rather than executing the offenders, a *panglima* might ceremonially draw blood from their thighs or upper arms using a knife or sword, razor blade, or piece of rattan. The *panglima* would then collect the blood in a plate and throw it into a river or the ocean, preventing the Tandayag from traveling into the uplands. According to one informant, during the 1960s, the former mayor Hadjes Asgali had ended executions and was then involved in less lethal punishments: "There were executions in the past. Bayuhan committed *sumbang* with his first cousin, so both of them were cut in the legs, but it was the Muslim [mayor] who did the punishment. His name is [former mayor] Asgali."

Panglimas enact these rituals alongside steep fines.[11] My informants generally expected offending parties to provide enough ceramic plates for each household potentially affected by the misfortune.[12] Often such fines represent a considerable financial burden, potentially requiring the provision of up to two hundred ceramic plates. The offending couple then gives these plates to *panglimas* of each hamlet, who will distribute them to each household. Community members break these plates and bury them in their swidden plots to shield agricultural production from environmental extremes. For most Pala'wan, the cutting and fining of offenders can generally mitigate the impacts of *sumbang* by fulfilling a ritual logic similar to that of execution: the removal of the essence or substance of sin from sites of Pala'wan habitation and the mollification of the Empu'. Previously, this took place in the discarding of offenders' bodies, but it now occurs through the transmission of their blood.

During 2011–12, three cases of *sumbang* among couples in two upland hamlets (two in Kapinpin, one in the hamlet of Mag-Agong) far into the forest interior were frequently discussed by Pala'wan people in chance meetings on forest trails, over meals, and in conversations about changing weather. These cases were a cause of particular anxiety among *panglimas* throughout the uplands, who saw themselves as responsible for ensuring the correct punishment for incest and maintaining normal climatic conditions. Each of these cases occurred between first cousins, generally the least serious of potential *sumbang* transgressions. The most intense scrutiny fell on an incestuous relationship that had developed in the distant hamlet of Mag-Agong. Two cousins had started an affair several months previously and were now living together. Further complicating the matter, it was commonly known that the woman was pregnant. Panglima Muku Tanduk explained:

> Because they are first cousin, maybe what happened is they drank alcohol and then did something stupid. . . . It means they must be separated because they are cousins. I can't decide on that matter. I told Meyreg to call both parties to meet in [the hamlet of] Bayabas to talk so they can agree on what do. If you can come, that will be on Friday. They can't be married because they are cousins, but the woman is already seven months pregnant. And the side of the man will have to pay the fines, and they must be separated.[13]

In line with the prohibition on executions, Muku and his son-in-law Panglima Meyreg of the hamlet Paratuong worked toward resolving the situation by meeting with the offending parties and their parents to fine and separate the couple and then draw blood from both offenders and offer it to the Tandayag. However, complicating matters, their parents refused to allow them to be cut. This resistance became a serious source of tension within the catchment for the following months. Other households in the uplands exerted pressure on Muku and Meyreg to resolve the issue and prevent further disaster from falling on the community. Ultimately, shortly before my departure in 2012, the death of the woman in childbirth settled this conflict. Meyreg later explained that the crime of incest must be punished, and without ritually cutting the couple, the woman's death was inevitable. The child itself, though a product of *sumbang*, was not viewed negatively—after all, my informants reasoned, it was not the fault of the child that its parents had sinned.

These cases of *sumbang* were not seen as isolated events or random acts of deviance but were situated within a broader perception of moral degeneration and, therefore, climatic decline. Though many Pala'wan people saw fines and bloodletting as reasonable substitutes for ritual executions, they also argued that without the threat of death, many young people no longer fear to commit incest and no longer respect the authority of customary leaders. Without punitive social sanction, incest has apparently now become common. Regardless of whether executions ever took place with the level of enthusiasm described in oral history or whether incest really has increased in frequency,[14] these perceptions form part of a social memory through which many *panglimas* understand present practices and climatic variation. These histories reveal that incest is believed to be increasing in frequency as Pala'wan lose the ability to control themselves and each other.[15]

STATE POWER AND PALA'WAN UNDERSTANDINGS OF CLIMATIC CHANGE

These accounts of incest and poor weather highlight how climate is understood by Pala'wan in explicitly moral terms; culpability for climatic disturbances is located within the sexual behavior of particular individuals and the inability of ritual specialists to enact punishment. However, a closer examination suggests this climate narrative is linked to larger and intersecting concerns surrounding livelihoods, forest resources, and the relationship between indigenous peoples and the state. Many Pala'wan reflect on local histories in which the execution of incest offenders was the most effective means of deterring immorality and ensuring proper environmental function. In doing so, they explicitly critique the Philippine government's restriction of customary authority and livelihood practices.

Many Pala'wan, regardless of their relationship to ongoing conservation efforts and other forms of state intervention in the uplands, draw direct connections between restrictions on customary punishments and climatic decline. A relatively young *panglima*, about thirty-five years of age, from a hamlet in the foothills of the *barangay* articulated this logic clearly:

> Before, according to the story of the old people, those who committed *sumbang* are sentenced with death, but right now we can't impose that punishment because that is against the law of the government. . . . That's why our tradition was changed, because of the government. Right now the only punishment is to bleed them and then use the plate and throw it into the sea, and then the sin will be forgiven . . . That's why, for me, it's much better if the old practices will be adopted today concerning that

issue, because if we think about it, it's your own flesh and blood, and then you have sexual relations. It is very immoral. It is not like planting rice, that you can eat whatever you have planted. We cannot do that in human relations. If God [Diyos] is merciful, he might not destroy this earth.

While within this man's lifetime the execution of incest offenders had never taken place, *panglimas* throughout the Inogbong uplands have maintained the importance of these practices in the absence of their performance. This importance was widespread, and many Pala'wan emphasized the connection between the state prohibition on execution and unfavorable weather throughout the *barangay* as part of attempts to understand extreme climatic variability. Though many *panglimas* have benefitted from an active participation in forest management projects (as this man had), the extent to which customary punishments could and should be practiced remained a point of contention. The perceived impotency of customary authority in the face of lowland state power is understandably a concern for traditional leaders despite the political and material incentives that arise from compliance with state policies. Among *panglimas* who reside in the foothills of the *barangay*, ambivalence surrounds discussions of the merit of customary executions. For example, one older *panglima* living in a hamlet adjacent to migrant paddy fields reflected on the inability to enforce punishments amid the rumors of incest circulating in 2011–12: "I'm trying to think of the new punishment for the *sumbang* because if you will talk to them [the offenders], they might not listen to you. Many people here committed *sumbang*, and we try to talk to them, but they do not obey. . . . And then Bada committed *sumbang*, but he won't agree to the punishment. So, right now, nobody follows the punishment for the *sumbang*." Nevertheless, he continued, "It's better these days, because before the people known to have committed *sumbang* were being killed without having any trial. But now you cannot just kill any person, even if that person is known to have committed *sumbang*."

However, among households more significantly impacted by increasingly restrictive aspects of forestry policy, discussions of *sumbang* contain far more explicit criticism of the Philippine state. Though articulated through the same cosmological principles, these ways of talking about incest and weather in the forest interior often frame a more pointed and active critique of state involvement in upland life. For example, a Pala'wan man, residing in the forest interior and relying heavily on swidden, complained of an inability to burn his plot for several years, linking climate change explicitly to a state ban on executing offenders of incest:

> Honestly speaking, it is the problem of the Pala'wan. The reason why we can't burn [our fields] is because . . . some people that are related to each other by blood are having a sexual relationship, like cousins. But before, during the time of our ancestors, if they know that such relationship exists, they look for the offenders, and then they will kill them. That's why we can't burn the swidden field. In Pinpin, they are cousins, because the mother of the boy and the father of the girl were siblings. Before, people are not like that. . . . Honestly speaking, there are more than three or four cases of *sumbang* here. . . .

Poorer households in the *barangay*'s more forested hinterlands also engaged in more explicitly politicized discussions of climate change. For example, a prominent *panglima* explained climate change by linking livelihood impacts to state policy in the uplands:

> The problem is the rain: we started to clear the field, and the rain is not that heavy, so we decided to continue clearing, hoping the weather will also be good, but suddenly the [heavy] rain comes. . . . For me, it is because of *sumbang* only—no other reason, because before, only one person committed *sumbang*, but now almost every year there is a *sumbang* case. . . . Because the government will not allow killing, the people are not afraid to commit *sumbang*. We can't do anything about the *sumbang* because the government forbids it. . . . Before, the government did not intervene with the customary punishment, and whoever committed *sumbang* will be killed right away, and then the body will be left under the sun. . . . That's why if ever we experience extreme heat and rain conditions, it is because of *sumbang*. So I think if it is possible to talk to the government, they should give the power to the customary leaders to impose certain punishment for committing *sumbang*, and because I don't have education, I can't do that. . . . That is also hard for me because I still have young children. But if not, I would fight for our rights. For example, you want to live a long time? How can you live a long time if there is excessive heat and rain?

Whereas Pala'wan closer to the lowlands are far more accepting, or at least resigned, to transformations of *sumbang* punishments and the decline of customary authority, these Pala'wan men in the forest interior reflect on older state practices (banning incest) to rhetorically resist their marginal position. The same forest-dependent households also draw connections between contemporary projects of environmental governance and climatic change through discussions of ceremonial decline.

Parallel to what was perceived to be rising incest in 2011–12 was the inability of ritual specialists to regulate the environment—an inability both implicitly and explicitly connected by marginal households to recent histories of swidden decline and environmental governance. Though there are a range of rituals that concern the mediation of climatic conditions, many Pala'wan people focused their anxiety on the large-scale sacrificial ceremonies—referred to as *ungsud* or *simaya*. Though the particulars may vary, the annual (or potentially less frequent) renewal of the agrocosmological order is a common theme of Pala'wan ritual life beyond Inogbong. For example, in the Kulbi-Kenipa'an catchment of the adjacent municipality of Rizal, it is known as *panngaris* and performed once every several years (Macdonald, 1997); in Barangay Rangsang, Rizal, as *simbung* (Novellino, 2002); and in the nearby *barangay* of Mainit, Brooke's Point, as *mundog* or *tambilaw* (Brown, 1991, p. 183).

Descriptions of these practices from Pala'wan in Inogbong roughly align with accounts of those conducted by indigenous peoples in other areas of the island. Rice wine (*tinapey*) is prepared and drunk, followed by days of prayer and dancing, during which requests for good weather and forgiveness for sins are made. Showing respect or directing platitudes or flattery toward the Master of Rice[16] will encourage this deity to relay these requests to the Empu' Banar, who will then halt the negative climatic conditions caused by the movements of the Tandayag. However, the community had not performed these large-scale sacrifices in over ten years prior to my arrival. Many Pala'wan saw this dereliction of ritual duty as contributing, alongside growing incestuous behavior, to poor climatic conditions. A Pala'wan man residing in the forest interior attributed ritual decline to increasingly poor agricultural production: "We can't do it anymore. Before, they can do some rituals because they have a good harvest, but now, how can we do it? There are still someone who does it, but is very rare because we don't have the rice to make [the rice wine]."

Like the rise of *sumbang*, ritual decline is also an extremely political phenomenon. Oral histories suggest that in the time of their ancestors, before the arrival of the Philippine state, when rice was plentiful, Pala'wan people in Inogbong performed such rituals whenever necessary. Despite these descriptions of a more abundant past, these sacrificial rituals are, and likely have always been, informed by a political economy of rice scarcity:

Producing one jar of rice wine requires approximately one fifty-kilogram sack of nonglutinous rice. This is a considerable, perhaps even overwhelming, amount of rice to dedicate to ritual purposes amid general perception of swidden decline. Such rituals also entail significant investments of time and labor to accommodate potentially hundreds of participants, who might journey from throughout and beyond Inogbong to attend. For households that rely more heavily on swidden production, the sacrifice of an entire sack of rice would most certainly result in hardship. These households lack the capacity to purchase surplus rice from the municipal market or migrant-run *sari-sari* stores with any regularity.

These statements regarding ritual practice are not separate from perceptions of social differentiation and of wider agricultural decline. In many cases, household members who understood themselves as marginalized by and from conservation efforts framed this decline in ritual practice in terms of poor harvests. These implicit connections are often made explicit; some households in the forest interior drew clear causality between the arrival of conservation projects and declining climatic conditions. One Palawan man explained this in terms of the presence of the PTFPP: "What I know is that when the PTFPP is not yet there, the weather conditions are balanced. There is enough rain and sun, but when the PTFPP is already here, the weather condition changed compared to the time of our ancestor, when there is no PTFPP."

Pala'wan people did not see deteriorating ability to produce or acquire rice as the only cause of ritual decline. Across interviews and everyday discussion surrounding weather, informants articulated the cultural and moral decline in the attitudes of others as a further reason why such rituals cannot, or will not, be performed. For example, other Pala'wan men explained poor weather and ritual decline in terms of cultural impurity:

> PALA'WAN MAN 1: Maybe it is because we seldom make rice wine now, and nowadays it is not pure, they mix it with gin. And maybe it is also because there is a change in the attitude of the people and their practices. Before, most of us wore a loincloth (*bahag*),[17] but now we use the modern clothing.
>
> PALA'WAN MAN 2: But for me, I am still wearing the *bahag* in my home.

Household members in the forest interior commonly held these views, and they implicitly directed them at Pala'wan who have become closer—culturally, economically, and spatially—to lowland Filipinos. Associations with the state and lowland economic activities, the instruments that criminalize and disempower swidden livelihoods, are seen to cause abandonment of icons of authenticity (for example, the *bahag* loincloth and rice wine) that now form the vocabulary of difference between indigenous people and lowlanders. Idioms of dilution and loss permeate these discussions of cultural transformation, which many Pala'wan see as related to wider socioecological function. These discourses of interconnected climatic and cultural change are reflective of finer-grained struggles at play *within* the catchment, as households that have been bypassed by the material benefits of lowland transformations and upland governance (in the form of conservation projects, such as the PTFPP) resist their marginality and the criminalization of their livelihoods. This resistance is performed by laying blame on government restrictions on swidden practices in addition to the cultural and moral decline among Pala'wan who associate with (and benefit from) conservation projects and state forestry policy.

While at first glance local climate discourses proximately place culpability on the growing immorality of Pala'wan behavior, these discussions also link state prohibitions on customary punishment to climatic decline. Though Pala'wan see the transformed ritual punishments as fulfilling a similar ritual logic within the confines of Philippine law, they also believe the ability of customary leaders to regulate *sumbang* is diminished without the threat of lethal social sanction. More marginalized households and *panglimas* in the forest interior strategically connect rumors of incest to impacts on swidden livelihoods and ultimately on their own well-being in order to explicitly critique state policy. The invocation of past practices, ones not witnessed directly, is an important commentary on the role of the Philippine state within Pala'wan life. By linking poor agricultural production and cultural degeneration (through the adoption of "modern" ways of being and doing) to participation in forestry projects, such households rhetorically critique both the contemporary forestry agenda of the Philippine state and the compliance of other Pala'wan households.

CONCLUSION

Indigenous people in Inogbong make sense of climate change through swidden practices as both a sensory interface with weather and a contested object of governance. Discussions of incest in the *barangay* in 2011–12 suggested that negative climatic change is the result of an increasing immoral society—that individual Pala'wan and their incestuous behavior are culpable for climatic changes and that their peers are unwilling or unable to commit material resources toward the ritual regulation of the environment. Ultimately, however, many Pala'wan link ritual decline and an associated rise in incest to the state's intrusion on upland life. Though for the Philippine state, the control of Pala'wan customary social practices, like ritual executions, is distinct in intent and practice from the governance of forest resources, for many Pala'wan, they are indistinguishable elements of a broader project of state control of live and livelihood.[18] In suggesting equivalence between "social" and "environmental" governance, such discourses of climate change reflect not only broader tension between indigenous people and the state but also the social differentiation within the community that has arisen in the past several decades. Social memories of a time before the arrival of the Philippine *gobyerno*, when weather was good, incest rare, and customary authority and tradition more certain, mediate contemporary relations between indigenous peoples and state actors. Local cosmological principles (for example, causal links between incest, the Tandayag, and customary punishment) are employed when reflecting on these idealized histories in the articulation of environmental transformations. The diverse causal connections subsequently drawn in statements regarding climatic change are, however, politically contingent and emergent or activated through emerging social differentiation in the uplands.

This is not to suggest that how Pala'wan understand both climatic and social transformation is crudely determined by material interest, shaped exclusively though relative involvement in and benefit from conservation projects and lowland economic opportunities. Climate change discourses reflect the genuine attempts of Pala'wan to understand issues such as cultural transformation, environmental uncertainty, and new geographies of social differentiation and state power. The reflection on histories of ritual violence and colonial practices features heavily throughout the uplands of Southeast Asia and is expressed not simply in resistance to contemporary practices of domination by nation-states but as a means of seeking forms of legitimacy in the past. Indeed, in Inogbong, climate knowledge speaks not only to environmental change but also to aspects of tradition and heritage among Pala'wan, who see their lives and lifeways as transforming. The point is not that political economy determines discourses of the environment, but that these densely textured discourses of climatic transformation cannot be fully understood divorced from struggles over the forest lives and livelihoods in which they are embedded and through which they are articulated.

Conclusion

Placing Blame

In June of 2011, during a break from fieldwork in Inogbong, I met the Pastor in the central Palawan municipality of Narra. I had traveled north of my primary field site to make some brief inquiries into a proposed forest management project aiming to cover the Victoria-Anepahan Mountain Range of central Palawan. Chatting with Tagbanua people living on the forest fringes, the Pastor and I quickly realized that few indigenous people in the area were aware of the project and its aims, despite the prominent project signage that greeted us as we traveled across the municipality on the Pastor's motorbike. Having already discussed the project with NGO staff in Puerto Princesa and reviewed some of the initial documentation, I thought, as did the largely disinterested indigenous participants, that much of the project was in many ways simply "more of the same." The project's staff emphasized to me that, as with the PTFPP and similar waves of community-based conservation initiatives that targeted indigenous people throughout Palawan Island over the past three decades, there would be no punitive restrictions on swidden—if Tagbanua people did not clear "old-growth" forest. The project would offer the usual suite of "alternative" livelihoods such as agroforestry, handicraft production, and other forms of market engagement.

Despite the repetition of older plans for indigenous lives and livelihoods that underpinned this project, it was distinct in that it tapped into a radical and exciting idea in regional forest governance: the hope that forest-based carbon trading could be a significant and even lucrative way to mitigate anthropogenic climate change. A consortium of local and regional nongovernmental organizations, with funding from the European Union, were working to implement a protected area that would be integrated into the

United Nations' "Reducing Emissions from Deforestation and Degradation" program. The Advancing the Development of the Victoria-Anepahan Communities and Ecosystems, or ADVANCE-REDD project, was part of a larger national effort at the time to ensure participation from the Philippines in future carbon-trading schemes and broader "REDD readiness" throughout the country. Like other climate change–focused projects, ADVANCE-REDD was dependent on a specific vision of indigenous peoples in relation to climate change. The Philippine strategy aimed to "respect indigenous knowledge" and was motivated by the externally produced "vulnerability" of the archipelago's inhabitants (Philippines REDD-Plus Strategy Team, 2010). This project, like REDD itself, was underpinned by a logic of scientific causality and moral culpability that seeks to address the historical carbon emissions by transferring resources to people whose lifeways can conserve forest resources.

When discussing the ADVANCE-REDD project with staff and targeted indigenous people in Narra, foremost in my mind was the following: What will happen when this project, or others like it, inevitably creep south to interface with the explanations for vulnerability and climate change that I had begun to encounter through my fieldwork in Inogbong? Pala'wan understandings of climate change, and their "mitigation" strategies in the form of ritual executions, do not sit easily alongside the scientific and moral foundations of REDD projects. Is this just another indigenous worldview that requires disciplining and reforming as part of forest governance on Palawan Island, or can these kinds of anxieties over colliding ontological assumptions speak to a larger relationship between climate change governance and indigeneity?

MOUNTAINS OF BLAME

Many anthropologists have grappled with self-blame and other unsettling indigenous climate narratives by placing them within their larger historical and political contexts. This book has focused on a specific project of marginalization: the ways in which indigenous peoples in Southeast Asia have been dispossessed and dislocated through colonial and postcolonial efforts to exploit forests, conserve biodiversity, and increasingly, mitigate climate change through preserving or enhancing stores of carbon. The state-owned forestlands of Southeast Asia and the Philippines, while present in the global imaginary as peripheral, have long been sites of interplay between local and global in which interlocking blame for environmental degradation motivates efforts to reshape upland livelihoods.

Colonial and postcolonial forestry efforts in the Philippines have been animated by a distorted and reductive understanding of indigenous peoples' agricultural practices. The production of environmental knowledge through state institutions has helped perpetuate a moral ecology of blame in which indigenous people, rather than industrialized logging or other forms of upland commodification, feature centrally in narratives of environmental decline. In contrast, ethnographic material from the Philippines and my own observations of Pala'wan swidden practices point to a vast gulf between officially authorized accounts of upland lives and how indigenous peoples might understand and value swidden agriculture. Since the 1970s, a growing focus in Philippine forestry on reforming and reshaping "livelihoods" rather than on disciplining *kaingin* overtly has operated as the interface for these contrasting views. Indigenous livelihoods emerged out of this period as the common grammar between targeted communities, policy makers, and project officers on the ground. However, this is not a rigidly deterministic process in which a singular "state vision" of life is imposed on pliant indigenous peoples. In the case of the PTFPP in Inogbong, newly institutionalized community-based forest management approaches gave the project a far-reaching mandate to comprehensively reform the way humans lived in upland areas, and it was animated by a politics that viewed issues of health, residency, and education as inseparable from conservation practice. At this unstable interface, simplified understandings of state and nonstate views were dissolved in the micropractices of environmental governance as the PTFPP ended and project rhetoric was taken up and refracted in the hands of customary leaders.

This is an analysis facilitated by conceptualizing livelihoods as contested points of spatial production, embodied in everyday practice. At the same time, this is an analysis informed not just by the generic potency of livelihood discourse and practice as sites of equivocation between indigenous and bureaucratic ontologies. Such a focus has a wider, abstracted analytical power, but the arguments I have constructed in this book are grounded in the material and spatial specificity of swidden agriculture itself. While swidden remains on the margins of many important arguments (or features as a case study) in critical environmental studies, it is a mode of economy and consumption that has not yet centrally contributed to social theory and academic imagination in ways that, for example, "hunting and gathering" or "animism" has (Nurit Bird-David, 1999). Yet, the interlinked contours of Philippine forest governance, indigeneity, and Pala'wan experiences of weather have been historically produced in relation to the specificities of swidden. This includes swidden's mobility, its reliance on fire, and its "integral" (in Conklin's [1957] terms) nature in many areas. Swidden is thus a particular a way of being and living, a livelihood, that links seemingly local concerns of indigenous cosmology, embodied experiences of weather, and environmental change directly into regional political economies of forest exploitation and global concerns over biodiversity.

This focus on swidden fields as plots of converging moral and practical concern, located in mountains of intersecting blame, provides complex rather than simplified narratives of Pala'wan experiences with weather that critically reexamine the way that indigenous peoples, their lives and bodies, are mobilized and targeted in efforts to solve problems of anthropogenic climate change. The comprehensive and intensive efforts to reshape indigenous livelihoods intrudes on, and complicates, seemingly straightforward narratives that might be applied to both understand and leverage Pala'wan experiences with weather. Observing conservation conflicts on the Italian island of Sardinia, anthropologist Tracey Heatherington (2010, p. 231) positions localized struggles within a "global dreamtime" of environmentalism, which she describes as a "jumbled set of narratives and visual images associated with environmental issues. They are systematically decontextualized from specific geopolitical contexts and tethered instead to markers of universal human experience and global citizenship. This fosters a sense of shared values, goals, and experiences that work to naturalize a post-national kind of 'imagined community.'"

The relationship between climate change and indigenous peoples is increasingly significant to this jumbled and shifting narrative assemblage. This position is often expressed in terms of "vulnerability" and "indigenous knowledge." Contextualizing unsettling Pala'wan experiences with weather both reveals and complicates the centrality of particular moral configurations of indigeneity to global environmental governance and collective human responses to climate change. What is this centrality based on, and what does its politics "do" in the world?

CLIMATE, VULNERABILITY, AND KNOWLEDGE IN THE GLOBAL DREAMTIME

The rise of "indigenous knowledge" and "vulnerability" as managerial categories in development practice has been well rehearsed and critiqued in the anthropological literature. Rather than seeing them as useful analytical concepts, critical scholars view both indigenous knowledge and vulnerability as discourses that reveal as much about those who deploy them as provide any kind of objective insight into the lives and experiences of indigenous peoples. Today, the primary locus of concern surrounding indigenous knowledge, for example, "is no longer the local knowledge systems as clearly separated 'there', but the hegemonic discourses that authorize essentialist representations of heterogeneous knowledges" (Nygren, 1999, p. 267). Vulnerability is similarly not only a technical means for understanding the impacts of disasters but a discourse that, "no less than the previous concepts of tropicality or development, also classifies certain regions or areas of the globe as more dangerous than others. It is still a paradigm for framing the world in such a way that it effectively divides it into two, between a zone where disasters occur regularly and one where they occur infrequently" (Bankoff, 2001, pp. 25–26). These simplifications are potent in the way they detract "attention and resources from underlying socio-economic inequalities that cause vulnerability in the first place" (Grove, 2018, p. 6).

Clearly, these are old and established though still lively debates. What value is there in launching yet another salvo against these conceptually bankrupt terms, already deployed only in the most careful or ironic fashion by critical scholars? Twenty years ago, Michael Dove (2000, p. 241) argued that indigenous knowledge operates fashionably in a cyclical pattern of reification and deconstruction, oscillating across the academy and development practice. In making this argument, Dove quotes Roy Ellen and Holly Harris's cutting aphorism that "indigenous knowledge is dead, long live indigenous knowledge." The simplifications arising from advocacy surrounding indigeneity in global climate change policy are a crest in a new wave of reification, rather than a distinctly new phenomenon.

The immense, though in some places contested, moral challenge of anthropogenic climate change means that older erasures and simplifications are reproduced, even by those alert to the problematic nature of the concepts through which indigenous experiences are understood and represented. These simplifications surrounding indigenous people have steadily become a central feature of, to borrow Heatherington's phrasing, the international "climate dreamtime" composed of academic research and global policy. For example, the documentation of the United Nations Framework Convention on Climate Change (UNFCCC) has increasingly rendered both indigenous environmental expertise (in the form of "knowledge") and victimhood (in the form of "vulnerability") not only useful but essential for combating climate change.[1] Intensive lobbying by indigenous groups and international NGOs prior to and during the 2015 UNFCCC 21st Conference of Parties meeting led to the prominent inclusion of indigenous knowledge in the resulting Paris Agreement. While the effectiveness of this movement has been downplayed or disparaged by its own participants as inadequate, the language of indigeneity and localness features heavily in the resulting discourse. The agreement notes that "parties acknowledge that adaptation action should follow a country driven, gender responsive, participatory and fully transparent approach, taking into consideration vulnerable groups, communities and ecosystems, and should be based on and guided by the best available science and, as appropriate, *traditional knowledge, knowledge of indigenous peoples and local knowledge systems*" (UNFFCCC, 2015, emphasis mine).

Outside of these formal frameworks, indigenous experiences with climate feature more prominently than ever in an ascendant global indigenous rights movement. Since 2011, for example, the globally influential environmental organization Conservation International has sponsored a series of "indigenous fellows" to showcase personal and community narratives of climate change and, in doing so, "elevat[e] their voices in the dialogue around climate resilience and conservation." In the form of written and video material available on the organization's website, indigenous peoples from throughout the world emphasize the impacts faced by their communities and how, in the words of one fellow, the solution to climate change "rests with indigenous peoples" (Conservation International, 2017). In one prominent blog post, Hindou Oumarou Ibrahim (2018), CI's senior indigenous fellow and noted climate activist, neatly defines this expansive vision of indigenous knowledge: "Indigenous peoples' traditional knowledge is the key to helping not only the world's indigenous peoples, but all of humanity, adapt and mitigate to climate change."

These efforts to cement a particular vision of indigeneity, both victim and savior, as a formal component global climate change policy and discourse are admirable. Like earlier movements in the 1980s and '90s to reorder the exclusion of indigenous peoples from international development and environmental governance, the promotion of varied knowledge traditions and experiences of suffering in climate change advocacy aims to challenge the ongoing marginalization of indigenous concerns and expertise from the global stage. In contrast to earlier movements, indigenous peoples, rather than white anthropologists or environmentalists (e.g., Brosius, 1997; Nadasdy, 2005), are often visibly at the forefront of these efforts. The prominence of articulate indigenous activists like Victoria Tauli-Corpuz and Hindou Oumarou Ibrahim makes it difficult to dismiss indigenous knowledge narratives as imposed Western romanticism. However, as important and worthwhile as these efforts are, these grandiose hopes might narrow the terms by which indigenous experiences with weather are understood. Like earlier environmental advocacy efforts, climate change activism often tends to discursively produce a homogenous "indigenous people" whose experiences might vary in the particulars but adhere to the broader narrative of the savior-victim complex (Brosius, 1997, p. 65). In reference to the "friction" of north-south environmental justice movements, Anna Tsing (2005, p. 13) notes that "allies rarely line up that well" and that there are a host of contingent sociopolitical practices that "smooth" such frictions. Indeed, these expectations pose a difficult question that is rarely reckoned with in either academic literature or policy circles: What happens if indigenous peoples do not, or cannot, meet these redemptive hopes or fit neatly into the "victim slot" (Hughes, 2013)?

In the context of the central role of climate change in REDD-Plus projects or the burgeoning move toward "climate smart agriculture," these are by no means hypothetical questions. The ADVANCE-REDD project on Palawan (though largely defunct) has prefigured today's regional governance landscape, where climate change mitigation and adaptation feature centrally in efforts to govern forests and their unruly inhabitants. For Pala'wan people, this entails a choice: be excluded from discussions surrounding climate change and full membership of the global indigenous rights movement, or reform their own narratives of climate change. Participation in prevailing trends of forest governance requires adherence to the temporal, moral, and causal logic of climate discourses, in which indigenous peoples are both victims and saviors in highly specific ways. This raises weighty questions beyond the scope of this book. Should, or can, the global indigenous rights movement reform to accommodate Pala'wan views, or should the Pala'wan reform themselves to access a global community of activism? The accounts of Pala'wan people hold the potential to undermine global indigenous rights movements by unpicking strategic essentialisms that produce animating disparities in culpability between industrialized nations and marginalized peoples. Rather than aiming to undermine already marginalized people, these kinds of reflections should further heighten attention to the implications of narrowed configurations of indigeneity in global discourse.

In addition to questions of policy and struggles for rights on the global stage, these expectations can shape the production of academic knowledge. Within this book, I have also positioned these debates to confront misalignments between my own positionality and expectations surrounding indigeneity, livelihood, and climate change. In place of simple and powerful narratives, I found more ambiguous stories of social and environmental change, and individuals enmeshed within complex moral situations. However, in gathering data on agricultural practices and perceptions of changing weather, I *could* have readily constructed a discussion that reifies narratives of indigenous expertise and vulnerability. Many indigenous people in Inogbong do indeed possess an intimate knowledge of forest environments, weather patterns, and agricultural practices, and they have deployed this knowledge in confronting uncertain weather. Much of this knowledge is articulated in reference to aspects of spirituality and mythology, which could have provided the material for constructing a romantic indigenous ecosubject. Similarly, the vulnerability of Pala'wan people to El Niño conditions could be easily explained as solely the product of a series of "easy targets"—the Department of Environment and Natural Resources, colonial forestry, or the state more generally. Creating this kind of story, however, would require papering over gaps, considerable variations, and misalignments between Pala'wan people's own explanations of climate and established climatological perspectives of changing weather in the region and my own. It would mean excising or ignoring troubling statements of culpability or desire for "nontraditional" ways of life or livelihood histories that shift away from "well-adapted" agricultural practices. Most significantly, it would require the careful elimination of accounts that unsettle the redemptive logic of indigenous knowledge or innocent vulnerability from the ethnographic record.

RESOLVING BLAME

But perhaps above and beyond these kinds of reflections, many scholars will largely be concerned with the following question: Does the contextualization of Pala'wan narratives and experiences in struggles over place unveil a hidden counterhegemonic potential? If self-blame can be reconfigured as metaphors of struggle and examples of subaltern agency, then it may not necessarily challenge existing narratives of blame and innocence that motivate action and, indeed, constitute global environmental struggles. The role of anthropologists in such scenarios would be the work of translating uncomfortable local rhetoric into a palatable explanation—this is Povinelli's "ethnographic magic." Science and Technology Studies scholar Sheila Jasanoff (2010, p. 238), writing on the vast epistemic gulf between embodied knowledge of weather and abstracted climate change science, remarks that "living creatively with climate change will require re-linking larger scales of scientific representation with smaller scales of social meaning."

These questions defy easy explanation and the ethnographic magic that can forge plausible links between the explanations of "the other" and global climate science. In this case, contextualizing accounts of self-blame within larger struggles over land and meaning can, in some respects, provide an alternative narrative that identifies an acceptable target of critique: the Philippine state's intrusive environmental interventions. Pala'wan people might proximately blame their own behavior, but digging deeper suggests that the target of these explanations for change is also larger material and discursive pressures on upland life. For many indigenous people, overt discussions of incest and ritualized executions are shaped in the context of larger concerns surrounding the state conservation measures on livelihoods. But does this tell us who is *really* being blamed here in any definitive sense? At the same time, Pala'wan people are, more than ever, vulnerable to the impacts of intense and recurring El Niño droughts. Indigenous people might be slowly but willingly forgoing swidden practices for more precarious forms of living, but how can these decisions be separated out from governmental intervention that has worked for decades to convince Pala'wan people to do so? State efforts to transform environmental subjectivities have shaped livelihood trajectories, but this process is bound up in the personal ambitions of customary leaders and a small but growing circle of Pala'wan people who have come to see a life outside of swidden. In this context, efforts to untangle the cause and effect, imposed or internally generated desires, and identify an ultimate "root cause" of vulnerability will lead to significant disappointment. These complications are therefore partial shifts, not complete resolutions. They raise more questions than they definitively answer, speaking to the analytical value that troubling accounts can provide in muddying and complicating the easy narratives dependent on particular configurations of indigeneity.

GLOSSARY

ABBREVIATIONS

C   Cebuano
P   Pala'wan
T   Tagalog

barangay (T) The lowest level of local political administration in the Philippines. On the southeast coast of Palawan Island, *barangays* typically align with river valleys and are often split between indigenous "upland" and a migrant-dominated flat coastal plain.
basakan (C) Irrigated paddy rice widely used by indigenous and nonindigenous people on Palawan Island
bisaya (P) A term used by Pala'wan people to refer to non-Muslim Filipino residents of Palawan Island. The term derives from the large number of migrants on Palawan originating from the Visayan Islands but is applied to all Christian migrants regardless of their actual origin.
carabao (T) Domesticated water buffalo common throughout the rural Philippines and highly valued as a draft animal
ganta (T) A unit of volume used to measure rice, equal to roughly two and a half kilograms of milled rice
kaingin (T) Rain-fed upland agriculture, whether practiced in a rotational fashion by indigenous peoples or as a pioneering strategy by migrant households
kaingiñero (T) A swidden farmer (pejorative, connoting environmental destruction and wastefulness)
panglima (P) The customary leader of a Pala'wan household cluster
sumbang (P) Incest
uma (P) Field or plot, largely used to refer to rotational swidden agriculture

NOTES

INTRODUCTION

1 The Arctic and small-island Pacific states, because of their sensitivity to early features of climate change and the striking visual imagery of environmental change that has emerged, have featured disproportionately in this literature (Krupnik & Jolly, 2001).
2 See the work of Dwyer and Minnegal (2000); Hughes (2017); and Cruikshank (2014) for a survey of the diversity of these efforts.
3 Well-known examples of this genre include Mark Elvin's (1998) exploration of imperial China's "moral meteorology." In a parallel sense, David Livingstone's (1999) work linking moral evaluation to tropical climates inverts these expectations to focus on Western colonial practices.
4 This includes, for example, the dissection of rural households as assemblages of capital (social, economic, natural, and so on), practices, and sociopolitical capabilities (Ellis, 2000).
5 Key literature that draws on, and expands, these Foucauldian perspectives includes the work of K. Sivaramakrishnan (1999), Arun Agrawal (2005), Tania Li (2007), and Pamela McElwee (2016).
6 A good example of this tension between punitive prosecution and educational efforts can be seen in the *Report of the Philippine Commission* for the year ending 1914: "Long experience seems to have demonstrated that the custom [*kaingin*] is too deep-seated to be uprooted by laborious processes of the criminal law, at least until a great deal of educational work has been done and a fuller understanding of the evil and its consequences has become more general. An active educational campaign on this subject and on the whole matter of conservation will be conducted during the year, in an effort to acquaint the great body of people throughout the islands with the benefits resulting from preservation of the forests and the harm which will occur if they are neglected or destroyed" (Denison, 1915, p. 115).

7 For a detailed overview of this process, see the doctoral theses of Katherine Warner (1979, pp. 79–98); Maria Lopez (1986, pp. 166–87); and Elaine Brown (1991, pp. 186–233); as well as James Eder's monograph *On the Road to Tribal Extinction* (1987).
8 While this process began much earlier throughout the Philippines under the Spanish and American colonial regimes, it was not until relatively late in the twentieth century that the bureaucratic infrastructure of forest management was fully exerted on much of Palawan Island.
9 These measures were interrupted by the Japanese occupation of the Philippines during the Second World War. However, following the cessation of hostilities, these efforts to police indigenous resource use were quickly resumed in some areas of the island. For example, in the early 1950s, the anthropologist Robert Fox (1954, p. 18) reported indigenous peoples in central Palawan were subject to increased measures to restrict the clearing of "valuable classes of wood" for agricultural purposes.
10 However, indigenous peoples of Palawan have been integrated for centuries into regional trading markets for ritual goods and important household items, and they have been subject to servitude and indebtedness for similar amounts of time. In addition to Chinese historical accounts, these perspectives are borne out by the observations of European explorers on Palawan during the late nineteenth and early twentieth centuries (Sawyer, 1900; Venturello, 1907; Whitehead, 1893).
11 Many, though by no means all, of these basic goods were produced for household consumption in swiddens. Tobacco and sugarcane, for example, were until recently widely cultivated in Pala'wan swiddens in Inogbong. Tobacco remains cultivated by households in the forest interior, whose members are poorly connected to cash economies.
12 For an overview of these processes, see the work of Melanie McDermott (2000, p. 160), Wolfram Dressler (2009), and James Eder (1987).
13 While many indigenous peoples have followed the example of pioneering migrants and moved into irrigated paddy rice (Conelly, 1983), opportunities to intensify and sedentarize rice production in the steep uplands of the islands are strictly limited by topography and water availability.
14 The ethnographic work of Wolfram Dressler (2009) and Melanie McDermott (2000) in and around the Puerto Princesa Subterranean River National Park in central Palawan Island has demonstrated the extremely fluid and shifting boundaries between indigenous peoples and migrant families.
15 Filipinos in the lowlands refer to indigenous peoples in Bataraza in a variety of ways. In the everyday discourse of lowland migrants, they are called *netibu* (native), tribal, or less frequently, "uncivilized." In local planning documents, Pala'wan are more likely to be referred to as "Indigenous Peoples" (IPs) or with the Tagalog term *katatubo*, or "innate"—a term that has wide currency throughout the Philippines in

official documentation but is little used on the ground in Bataraza, even among local government employees.

CHAPTER ONE: MAKING *UMA*, IMAGINING *KAINGIN*

1 I was a contributor to an op-ed piece that responded to the framing and publication of this article, which appeared in the *Philippine Daily Inquirer* later that year.
2 There are few considerations of *kaingin* in Philippine culture as anything other than backward and destructive. One poignant counterexample that offers a vision of an alternative trajectory in which *kaingin* could be an emblem of heritage rather than of primitive wastefulness comes from the work of Philippine poet and writer Nestor Gonzalez, whose depictions of rural life include vivid and nostalgic descriptions of *kaingiñero* families without castigation. See, for example, the collected volume of Nestor Gonzalez's writings *Children of the Ash-Covered Loam and Other Stories* (Gonzalez, 1954).
3 From its inception in 1848, Spanish state forestry (and subsequently colonial forestry) was strongly influenced by the German approaches to scientific forestry. As elsewhere in Southeast Asia, forestry officials of both Spanish and Filipino origins were extensively trained in German and French traditions of forest management (Bankoff, 2011).
4 The exclusion of indigenous perspectives is readily seen in early American assessments and studies of forest resources by colonial scientists and foresters such as Elmer Merrill (1908), Harry Whitford (1911), and William Brown (1919). Absent, entirely, from these accounts is the subsistence value that forests held through swidden cultivation or the collection of nontimber forest products.
5 Early regrowth after fire is highly favored by cattle, while older cogon is generally unpalatable, hence the need for regular burning to provide fodder (Dove, 1997).
6 The American colonial administration was, for example, particularly keen on quickly mechanizing the Philippine logging industry.
7 The colonial photography of Dean Worcester has been extensively dissected for its representation of indigenous peoples (Vergara, 1995). The influence of early forestry photography has yet to be analyzed to the same degree.
8 Eventually Filipino foresters educated in the US would come to direct the Bureau of Forestry itself; the first Filipino director of forestry, Florencio Tamesis, graduated from the University of Washington with a postgraduate degree in forestry (Nano, 1951, p. 81).
9 Postwar anthropological research by Filipino scholars continued to stubbornly rehearse colonial-era narratives. In a striking example, a

study completed by a Filipino anthropologist in the 1960s concluded that "this group and ... other similar ethnic groups in this country ... are still practicing the so-called *kaingin* system of farming and may be considered, as such, one of the weakest links in Philippine economy and society today" (Lopez, 1968, pp. 1–2).

10 A series of presidential decrees and administrative orders in the 1970s and '80s granted tenurial recognition to residents on some state-owned lands, on the conditions, however, "that the area is developed, planted with agricultural crops using effective erosion control practices like terracing, and that there are basic structures like schools and churches clearly existing" (Molintas, 2004, pp. 286–87). Even today, most indigenous communities on Palawan Island would not satisfy these criteria.

11 More communal tenurial mechanisms were contingent on the development of agroforestry plantations and the limitation of new plots—and were still aimed at instilling a sense of individual ownership in land, albeit at the discretion of the community. Regardless, community contracts were relatively rare in practice.

12 Harold Olofson (1980, p. 70) notes that the Tagalog word *kaingin* is often used to refer to all upland agriculture. Many *kaingiñeros* in the Philippines are, in fact, permanent upland farmers, and the use of the term *kaingin* is an imposition of forestry officials (Kummer, 1992). As a result, the creation of meaningful categories of upland agriculture has proven difficult. In contrast, the Pala'wan term *uma* (literally meaning "plot") encompasses a wide range of rain-fed, upland agriculture (i.e., a fixed and/or ploughed plot). Among Pala'wan, the *uma* is contrasted to paddy rice fields (*basakan*, a Cebuano term commonly used on Palawan) and commercially oriented cash crop plantations.

13 In conversation, many Pala'wan would use the term *kaingin* for my convenience. Among themselves, however, *uma* is the preferred term.

14 Many researchers who have worked among indigenous peoples on Palawan have framed similar questions of personhood in terms of swidden cultivation. Katherine Warner (1979, p. 8), for example, notes that what made her a "social being" during her fieldwork among neighboring Tagbanua peoples was the establishment of a household and the preparation of a swidden plot. In contrast, Maria Lopez (1986, p. ix) suggests that what made her "acceptable" to the Pala'wan households with whom she worked was the fact that she did not make a swidden in a region with a fast-disappearing land frontier. Whether or not social acceptability is granted by participating in agricultural practices, what these accounts highlight is how *uma* making is undertaken by and defines newly formed households as independent agricultural units and social entities within broader hamlet and catchment-level relations.

15 Field sizes are typically estimated in increments of half a hectare.

16 Palawan's province-wide system of environmental zoning is simplified on the ground by Pala'wan and lowland *barangay* officials; clearing *malalaking kahoy* or "big trees" is illegal and often results in imprisonment or fines for offenders.

17 Anthropologists have highlighted how these beliefs have been vital in shaping the mosaiclike structure of Palawan's forests, as spiritual beings were thought to be present in certain trees or areas ("sacred groves") (Fox, 1954; Olofson, 1995).

18 "Brushland" is an official land use category (a disturbed forest area dominated by small shrub or brush) that has entered the common vernacular of lowland Filipinos on Palawan Island.

19 Many of my informants suggested that the best or "standard" time for burning is the first Friday of March.

20 These moments of collective labor are also sites of play—charcoal is smeared on participants to break the tension of an arduous planting schedule and introduce an element of joy into labor. The blackening charcoal is also said to be symbolic, representing the darkened clouds that bring forth the southwest monsoon.

21 More eclectic choices that were once common are not included. Crops such as sugarcane were said to be once-essential components of any swidden field but are now planted by only a few households in the upper portion of the *barangay*.

22 The categorization of unseen entities into benevolent and malevolent is common in descriptions of Pala'wan spirituality. In practice, these relations are somewhat more ambivalent.

23 Empu' and Diyos are often used interchangeably by nominally Christian Pala'wan. In such syncretic households, the indigenous deity who sits at the head of the semianimistic pantheon of divinities, Empu', is simply called God/Diyos without radical structural transformations in cosmology. However, as with *kaingin*, in conversation many Pala'wan would translate Empu' as Diyos for my convenience. The use of Diyos should therefore not necessarily be interpreted as a marker of strong Christian convictions.

24 While elements of cosmology may be similar for different Pala'wan groups, there is significant variation in ritual behavior, spiritual causality, and the names and functions ascribed to various deities across (and within) catchment areas.

25 In Inogbong, it is difficult to talk of individual or groups of Pala'wan as definitively Christian or not. Many indigenous people who have had extensive contact with missionaries may identify as Christian in that they suggest they attend church services. In practice, however, they never attend church and continue to perform indigenous rituals. Often, Christian and indigenous cosmological precepts are blended together.

Furthermore, in other Pala'wan communities, religious identity is characterized by a high degree of fluidity across the lifespan of individual men and women (Macdonald, 1993).

26 *Simaya* and *ungsud* are broad terms that refer to a wide range of ritual and interpersonal actions, encompassing, for example, both small-scale offerings and the contribution of bride wealth.

27 Though cared for or owned by the distinctly nonhuman Empu't Paray, the *diwata* does not reside within and provide the vitality of the rice itself. Rather, through the sacrifice of a human child, rice is imbued with a soul (*kurudwa*). The ascription of a kind of personhood (*taw*) to rice refers to both the role of the Empu't Paray and the notion that rice possesses humanlike affect that impacts its physical growth.

28 Honey is a substance of similar potency. It is the substance that has the same ritual capacity as rice, though it is far less frequently used and discussed in the same capacity.

29 Beyond swidden agriculture, ritual and ceremonial life focuses on health (or illness), another key anxiety not disconnected from swidden practice among Pala'wan people.

30 These exchanges between humans and *diwata* are ideally seen as reciprocal among Pala'wan (Revel, 1998) and emphasize balance between good and evil cosmological forces (Macdonald, 1992).

31 A parallel ritual may take place when honey wine (*simbug*) is used in place of *tinapey*. However, the substance used to call forth the *diwata* is not contingent on the type of request; rather, it alters the pathway of communication, as Empu't Burak (the Master of Flowers, guardian of bees) instead of Empu't Paray is called forth to mediate requests for aid when honey wine is used.

32 These kinds of statements should not necessarily be taken as an objective account of ritual decline. Other anthropological accounts of other Pala'wan groups suggest that analogous rituals have typically been performed only at long, irregular intervals (Macdonald, 1997).

CHAPTER TWO: ROOTED IN PLACE

1 Southern Palawan was and continues to be home to significant numbers of Muslim lowlanders (Eder, 2010). The American anthropologist Henry Beyer (1917) estimated the "Moro" population of Palawan in the early twentieth century at 20,000, presumably concentrated largely in the south of the island.

2 Within national environmental discourses in the Philippines, the uplands are often considered spaces dominated by forests, while the lowlands are areas of agriculture; however, this stark distinction is relatively recent. For example, lowlands now almost completely

dominated by agriculture were once heavily forested, as early accounts by naturalist John Whitehead (1893) attest. The rise of paddy farming, along with copra production, is complicit in the physical crafting of lowland landscapes on Palawan by Christian Filipino settlers.

3 Worcester (1913, p. 81) notes that it was part of US colonial policy to encourage sedentary farming among the Tausug of southern Palawan by distributing ploughs and other agricultural tools as a means of dissuading them from piracy. The logic of this intervention was that settled populations of Moros might be more easily controlled by American military forces (Ocampo, 1996, p. 36). French explorer Frederic Sawyer (1900, p. 318) notes that by the turn of the century, the Moros of southern Palawan had already begun to take up agriculture as an alternative means of livelihood but that all laborious tasks were "performed by slaves."

4 With the partial exception of rice, how these items became essential for indigenous household reproduction (as is the case today) is therefore not politically neutral or natural but situated within the exploitation of indigenous land and labor on Palawan. The stories of migrants and indigenous elders speak to how the desire for these goods was actively cultivated by pioneering farmers and then later leveraged in exploitative debt relations to expropriate land from indigenous people.

5 Working in the thick mud and water of the paddy fields was considered unclean and a potential source of skin disease by indigenous households. As such, work in migrant paddy fields was initially avoided due to health concerns.

6 This legal concept refers to the colonial policy that vested authority to dispense title solely with the Crown.

7 This mandate was reproduced by the Philippine national government in the 1987 constitution, which asserts that "with the exception of agricultural lands, all other natural resources shall not be alienated [from] the full control and supervision of the State" (Constitutional Commission, Article XII).

8 This process began as early as 1929 with the arrival of the first "land classification team" (Brown, 1991, p. 88), though in practice much of Bataraza remained surveyed until the 1950s and beyond.

9 Language was a common barrier to land reform elsewhere in the Philippines, as poor and marginal tenant farmers often had little command over Tagalog or English, the languages of administration (Kwiatkowski, 1998, p. 42).

10 Oral histories generated with my interlocutors in Inogbong speak to processes of dispossession that mirror those in accounts told to researchers Maria Lopez (1986) and Elaine Brown (1991), who also worked with Pala'wan people on the east coast.

11 A second phase of the PIADP began in 1990, focusing on northern Palawan, and operated parallel to the PTFPP, which focused on the south of the island (Asian Development Bank, 2002).
12 Efforts to sedentarize and commercialize indigenous peoples' livelihood practices sit within a long history of forest management in the Philippines but were also part of a provincially specific discussion of the role of *kaingin* within Palawan forests. For example, during workshops held with indigenous peoples on Palawan in 1996 to formulate a response to new Ancestral Domain legislation, it was agreed that "energy should be invested in the development of best practices for shifting cultivation in the area currently under a *kaingin* rotation cycle towards a gradual increase of permanent agriculture (fruit trees, gardening). Moreover, increased income derived from non-timber forest product resources would decrease dependence on *kaingin* crops for daily needs" (de Beer & McDermott, 1999, p. 157).
13 The distribution of *lakatan*-variety bananas and *kalamansi* had perhaps the greatest lasting impact, and these crops now form an important source of cash income for many households in the foothills. Mango, pepper, and cashew trees were distributed with less commercial success.
14 Due to lack of familiarity with Tagalog and lowland social etiquette, many Pala'wan in the forest interior have difficulty negotiating with migrants and are in some cases paid only half the daily wage of more experienced Pala'wan in the foothills.
15 In contrast to swidden, in which planning is often shared between husbands and wives, paddy rice is performed largely by men in upland Inogbong. The role of Pala'wan women in these activities was far less visible to me.
16 Under the Local Government Code (Republic Act No. 7160), the *barangay* and municipality (both considered "Local Government Units") are charged with environmental enforcement (in addition to the provision of other basic social services) in exchange for an "Internal Revenue Allotment" from the national government. *Barangay* and municipal officials are often invited to seminars organized by the Palawan Council for Sustainable Development (and nongovernmental organizations), where they are educated about provincial environmental regulations and their responsibilities in enforcing them.
17 For Pala'wan people and other indigenous groups on the island, the fear of imprisonment is significant and often disproportionate to its likelihood; the threat of separation from land and family is a powerful deterrent to illegal activity, even based on rare or spurious accounts. W. Thomas Conelly (1996, p. 207), for example, describes the impact of arrests among Napsaan Tagbanua swidden agriculturalists on controlling cultivation in forested areas.

18 During my fieldwork, the family was in the process of subdividing and retitling their lands among Peminta's children. Burit wanted the security of legal title in order to develop the land on an individual basis.

CHAPTER THREE: INSIDIOUS VULNERABILITIES

1 As a partial exception, Cynthia Fowler (2012) weaves local considerations of the 1997–98 event throughout her exploration of Kodi fire use.
2 The rapid alternation between extreme El Niño and La Niña phases has been associated for centuries with catastrophic drought-flood dynamics throughout the archipelago (Warren, 2013).
3 The Marangas River is significantly larger than the Inogbong River, hence its use as a marker of extreme drought.
4 The forest fires on Palawan during 1997–98 were infamous for their destruction of what was coming to be seen as the last remaining forest frontier of the nation, which put significant pressure on local authorities to abate the blazes.
5 The clearing of firebreaks in swidden burning (*gahit*) was an important part of PTFPP information dissemination campaigns.
6 Debt was not an important coping mechanism. As noted previously, Pala'wan are either unwilling or unable to access formal credit markets (bank loans) and informal credit markets (making *utang* with *sari-sari* stores).
7 The use of wild yams during droughts is common throughout Palawan Island and the Philippines. See, for example, Dario Novellino (2002).
8 Nevertheless, comparative historical research throughout Southeast Asia would suggest that starvation in forested environments—as opposed to hunger—was infrequent due to the wide availability of food within forest landscapes (Henley, 2005, pp. 359–60). This does not preclude, of course, pervasive malnutrition and associated morbidity.
9 Informants recalled a range of prices between 90 to 120 pesos per *ganta*—in some cases more than double the usual cost in 2011–12. This rice was of poor quality and was perceived by both indigenous people and migrant lowlanders as possessing an unpleasant aroma that made it very undesirable.
10 This is a practice that has its origins in early small-scale logging on the island and continues into the present, with migrant loggers paying indigenous people to guide them to and store harvested timber—a practice that also effectively transplants much of the risk to Pala'wan. During my fieldwork, at least one Pala'wan man was apprehended and threatened with monetary fines and the possibility of jail for storing *ipil* (*Intsia bijuga*) boards for illegal loggers.

11 Jill Belsky and Stephen Siebert (1983) note that during the 1982–83 El Niño–induced drought in Leyte, all rain-fed agriculture experienced a total crop failure except long-fallowed swiddens in the forest interior.
12 Though some rice is said to have survived in the uplands, it is unlikely to have been eaten or traded. Beyond the need to preserve seed stock for the following year, as Macdonald (2007, pp. 53–54) observes, the free exchange of rice among Pala'wan rarely occurs outside the immediate family or hamlet—more extended kin living in more distant areas, however, have greater social claim to root crops, which are often freely given in times of need.
13 It has been argued that these patterns of employment in southern Palawan encourage indigenous farmers to invest less in swidden agriculture and forest production extraction (Shively, 2001).
14 Perhaps more so as transitions into paddy rice are associated with abandoning swidden entirely, in contrast to those specializing in wage labor, who will generally (though not always) retain root crop–focused *uma* plots. Even when swiddens are retained by paddy rice farmers, they tend to be smaller and less well tended and contain less crop diversity. Such households do possess capital assets that may be liquidated in times of need, such as carabao or hand tractors. While such coping strategies may provision for immediate needs, they are ultimately detrimental to future production.
15 In contemporary practice, cassava is often "banked" in older swidden fields to be drawn on in times of hardship. While root crops such as cassava have likely long been planted in excess, there is now at least a rhetorical emphasis on the strategic element of this practice. Cassava can remain edible up to a year after maturation, though the texture and taste decline markedly and become woody and unpalatable.

CHAPTER FOUR: EL NIÑO AND INCEST

1 The representation of local views as explicit barriers to adaptation is widespread in more climate change studies. See, for example, Anja Byg and Jan Salick (2009) and Shalini Lata and Patrick Nunn (2012).
2 While other aspects of the weather are critical to the productivity of swidden plots—for example, the intensity of heat, cyclonic storm activity, or direction of wind—discussions of climate focus primarily on seasonality in terms of the presence or absence of rain. Similarly, weather is a component of varying significance and sensitivity for many livelihood activities (e.g., collecting *almaciga* is made more difficult by wet, muddy trails in the rainy season).
3 See Nicole Revel (1990) for a comprehensive list of "traditional" Pala'wan uses of constellations and their significance—the markers I have listed

are those that retain significant meaning to my informants, however divorced they may be from past or idealized understandings of indigenous knowledge.

4 March is a significant month in burning schedules. If a significant amount of vegetation is being dried in a plot and rain falls in March, it is unlikely to dry again in time for burning in late March or early to mid-April. This is simply one indicator, and similar variability is apparent if other indicators are examined, for example, rainfall the following months.

5 Some Pala'wan claim that the proximity and position of Venus (*buntatala*) relative to the moon not only signals that incest has occurred but also indicates what kind of incest has occurred—thereby predicting not only climatic perturbations but also their specific severity. However, this is a rather indirect prediction of weather, and it primarily signals moral transgression. Revel (1990) also suggests that in nearby catchment areas, slight variations in the star rise and fall have predictive capability. While Benjamin Orlove et al. (2000) note that similar observations elsewhere in the world are linked to the ability to predict ENSO events, such connections are not locally drawn in Inogbong.

6 As a former *barangay* official noted of climatic uncertainty, this variability is a greater problem for swidden livelihoods than for the largely irrigated paddy rice farming in the lowlands: "I think only the Pala'wan people are affected by this because the people here are continuously doing *basakan*, but the Pala'wan up there can't burn their *kaingin*."

7 Incest has long been, and remains, a critical concern for many indigenous groups on Palawan Island that consistently features in ritual and ceremonial practices and wider mythology.

8 Closer pairings entail more severe environmental consequences. For example, pairings between first cousins are generally seen as the most common form of sexual transgression and, though causing serious environmental imbalance, are considered relatively mild when compared to sexual relations between more immediate relatives that cut across generational lines (such as father and daughter), which threaten the world with complete destruction.

9 For example, the young not listening to the old, various traditions being abandoned, and increasing selfishness among Pala'wan were other explanations were also put forward as reasons for a progressively deteriorating climate.

10 The sensory aspects of incestuous practices, for example, the smell and heat generated by incest, are also present in other Pala'wan communities (Macdonald, 1997).

11 The use of fines in regulating social behavior pervades all aspects of Pala'wan life. For example, transgressions in customary marriage, such

as defying matrilocal norms when first married, are permitted when small monetary compensation is made to the bride's parents.

12 The impact of incest is geographically confined to the area in which the sins have been committed, for example, within and adjacent to a river valley, but not extending over the entirety of southern Palawan or over the mountain range.

13 Here, Panglima Muku links moral degeneration in the form of drinking to incest. When asked about the role of liquor, he stated, "Yes, that's the root cause of the *sumbang*."

14 There are no data to evaluate changes in the frequency of incestuous relationships over time, either in Inogbong or among Pala'wan groups. Given the prevalence of incestuous themes in the folklore and mythology of Pala'wan and related indigenous groups on the island, I suggest that intense anxiety surrounding incest has always been a feature of social and political life and is now being leveraged in new areas of contestation.

15 Interestingly, Charles Macdonald (2007, p. 78) calculated in 1989 that 1.4 percent of unions in Kulbi-Kenipa'an, in the neighboring municipality of Rizal, were, using the criteria noted above, incestuous.

16 The focal point of such rituals varies in accordance with the substance used to create the alcohol. Rice is used in ceremonies directed toward the Master of Rice, who is seen to ultimately transmit requests to Empu' Banar. However, if honey is used, then the requests are directed toward Empu't Barak, the Master of Flowers. Though honey, like rice, is a divine gift and therefore an acceptable sacrifice to command the attention of Empu' Banar, rice is given preference or priority in discussions of ritual practice.

17 The *bahag* is a loincloth closely associated throughout Palawan with "authentic" or "traditional" indigenous people and behavior.

18 This has been most recently manifest in "4Ps" (Pantawid Pamilyang Pilipino Program), which delivers cash to poorer Filipino households contingent on their children's school attendance, use of regular health checks, and so on. While not specifically targeting indigenous households, such programs, at least locally in Bataraza, are clearly believed to be transforming what the state sees as the backwardness and irresponsibility of indigenous peoples by financially encouraging them to conform to mainstream notions of hygiene, health, and education.

CONCLUSION: PLACING BLAME

1 See the work of James Ford et al. (2016) for a comprehensive overview of references to indigenous peoples and knowledge in the varied outputs from the United Nations Framework Convention on Climate Change.

BIBLIOGRAPHY

Agrawal, Arun. 2005. *Environmentality: Technologies of Government and the Making of Subjects*. Durham, NC: Duke University Press.

Ahern, George. 1902. *Report of the Chief of the Forestry Bureau for the Period from July 1, 1901, to September 1, 1902*. Washington, DC: Bureau of Insular Affairs.

———. 1910. *Annual Report of the Director of Forestry of the Philippine Islands for the Period July 1, 1908 to June 30 1909*. Manila: Bureau of Forestry.

———. 1911. *Annual Report of the Director of Forestry of the Philippine Islands for the Period July 1, 1909 to June 30 1910*. Manila: Bureau of Forestry.

Akramgolteb. 2015. Comment on Summer Not All Beach in Palawan; It Is the Season to Burn Forests. *Philippine Daily Inquirer*. Retrieved from: https://newsinfo.inquirer.net/684378/summer-not-all-beach-in-palawan-it-is-the-season-to-burn-forests/2.

Allan, Robert J. 2000. Enso and Climatic Variability in the Last 150 Years. In F. Diaz & V. Markgraf (Eds.), *El Nino and the Southern Oscillation: Multiscale Variability and Its Impacts on Natural Ecosystems and Society* (pp. 3–55). Cambridge: Cambridge University Press.

Anonymous. 1931. The Kaingin Menace. *Makiling Echo*, *10*, 2–3.

Asian Development Bank. 1991. *Project Completion Report on the Palawan Integrated Area Development Project in the Philippines*. Manila: Asian Development Bank.

———. 2002. *Project Performance Audit Report on the Second Palawan Integrated Area Development Project in the Philippines*. Manila: Asian Development Bank.

Astuti, Rita. 2017. Taking People Seriously. *HAU: Journal of Ethnographic Theory*, *7*(1), 105–22.

Bankoff, Greg. 2001. Rendering the World Unsafe: "Vulnerability" as Western Discourse. *Disasters*, *25*(1), 19–35.

———. 2007. One Island Too Many: Reappraising the Extent of Deforestation in the Philippines Prior to 1946. *Journal of Historical Geography*, 33(2), 314–34.

———. 2009. Breaking New Ground? Gifford Pinchot and the Birth of "Empire Forestry" in the Philippines, 1900–1905. *Environment and History*, 15(3), 369–93.

———. 2011. The Science of Nature and Nature of Science in the Spanish and American Philippines. In D. F. Ax, N. Brimnes, N. T. Jensen, & K. Oslund (Eds.), *Cultivating the Colonies: Colonial States and Their Environmental Legacies* (pp. 80–93). Columbus: Ohio University Press.

Bellwood, Peter. 2006. Asian Farming Diasporas? Agriculture, Language, and Genes in China and Southeast Asia. In M. T. Stark (Ed.), *Archaeology of Asia* (pp. 96–118). Oxford: Blackwell.

Belsky, Jill M., & Siebert, Stephen F. 1983. Household Responses to Drought in Two Subsistence Leyte Villages. *Philippine Quarterly of Culture and Society*, 11(4), 237–56.

Bensel, Terrence. 2005. Climate Forecasts and Government Response: The Case of the 1997–1998 El Niño Drought Event in the Philippines. *Philippine Geographical Journal*, 49(1–4), 86–107.

Bessire, Lucas, & Bond, David. 2014. Ontological Anthropology and the Deferral of Critique. *American Ethnologist*, 41(3), 440–56.

Beyer, Henry O. 1917. *Population of the Philippine Islands in 1916*. Manila: Philippine Education.

Biersack, Aletta. 2006. Reimagining Political Ecology: Culture/Power/History/Nature. In A. Biersack & J. B. Greenberg (Eds.), *Reimagining Political Ecology* (pp. 3–42). Durham, NC: Duke University Press.

Bird, Michael I., Boobyer, Ella M., Bryant, Charlotte, Lewis, Helen A., Paz, Victor, & Stephens, W. Edryd. 2007. A Long Record of Environmental Change from Bat Guano Deposits in Makangit Cave, Palawan, Philippines. *Earth and Environmental Science Transactions of the Royal Society of Edinburgh*, 98(1), 59–69.

Bird-David, Nurit. 1999. "Animism" Revisited: Personhood, Environment, and Relational Epistemology. *Current Anthropology*, 40(S1), S67–S91.

Broad, Robin, & Cavanagh, John. 1993. *Plundering Paradise: The Struggle for the Environment in the Philippines*. Berkeley: University of California Press.

Brosius, Peter J. 1997. Endangered Forest, Endangered People: Environmentalist Representations of Indigenous Knowledge. *Human Ecology*, 25(1), 47–69.

———. 2006. What Counts as Local Knowledge in Global Environmental Assessments and Conventions? In W. Reid, F. Berkes, T. Wilbanks, & D. Capistrano (Eds.), *Bridging Scales and Knowledge Systems: Concepts and Applications in Ecosystem Assessment* (pp. 127–44). Washington, DC: Island Press.

Brosius, Peter J., Tsing, Anna L., & Zerner, Charles. 2005. *Communities and Conservation: Histories and Politics of Community-Based Natural Resource Management*. Walnut Creek, CA: Altamira.

Brown, Elaine C. 1991. *Tribal Peoples and Land Settlement: The Effects of Philippine Capitalist Development on the Palawan*. PhD thesis, State University of New York.

Brown, William H. 1919. *Vegetation of Philippine Mountains: The Relation between the Environment and Physical Types at Different Altitudes*. Manila: Department of Agriculture and Natural Resources.

Bryant, Raymond L. 1997. *The Political Ecology of Forestry in Burma*. Honolulu: University of Hawaii Press.

———. 2008. *Nongovernmental Organizations in Environmental Struggles: Politics and the Making of Moral Capital in the Philippines*. New Haven, CT: Yale University Press.

Bulloch, Hannah. 2017. *In Pursuit of Progress: Narratives of Development on a Philippine Island*. Manoa: University of Hawaii Press.

Byg, Anja, & Salick, Jan. 2009. Local Perspectives on a Global Phenomenon—Climate Change in Eastern Tibetan Villages. *Global Environmental Change*, 19(2), 156–66.

Cadeliña, Rowe V. 1986. Social Forestry in the Philippines' Uplands: A University Perspective. In Y. S. Rao, M. W. Hoskins, N. T. Vergara, & C. P. Castro (Eds.), *Community Forestry: Lessons from Case Studies in Asia and the Pacific* (pp. 103–34). Bangkok: Food and Agriculture Organisation.

Calanog, Lope A. 1977. The Government Kaingin Management Project. *Canopy*, 3(3), 6–7.

Condominas, George. 1957. *We Have Eaten the Forest: The Story of a Montagnard Village in the Central Highlands of Vietnam*. New York: Hill and Wang.

Conelly, W. Thomas. 1983. *Upland Development in the Tropics: Alternative Economic Strategies in a Philippine Frontier Community*. PhD thesis, University of California at Santa Barbara.

———. 1996. Agricultural Intensification in a Philippine Frontier Community: Impact on Labor Efficiency and Farm Diversity. In D. G. Bates & S. H. Lees (Eds.), *Case Studies in Human Ecology* (pp. 289–310). Boston: Springer US.

Conklin, Harold. 1957. *Hanunoo Agriculture: A Report on an Integral System of Shifting Cultivation in the Philippines*. Rome: Food and Agriculture Organisation.

Conservation International. 2017. *Floods & Climate Change Spark Indigenous Knowledge in Peru*. Conservation International. Retrieved from: www

.youtube.com/watch?time_continue=13&v=r1qF8aMBaBQ&feature=emb_logo.

Constitutional Commission. 1987. *The 1987 Constitution of the Republic of the Philippines*. Retrieved from: www.chanrobles.com/philsupremelaw2.html.

Craggs, Dennis. 1998. *Agriculture and Livelihood*. Puerto Princesa: PCSDS.

Cramb, Rob. 2000. *Soil Conservation Technologies for Smallholder Farming Systems in the Philippine Uplands: A Socioeconomic Evaluation*. Canberra: Australian Centre for International Agricultural Research.

———. 2007. *Land and Longhouse: Agrarian Transformation in the Uplands of Sarawak*. Copenhagen: NIAS Press.

Cruikshank, Julie. 2014. *Do Glaciers Listen? Local Knowledge, Colonial Encounters, and Social Imagination*. Vancouver: University of British Columbia Press.

Dacanay, Placido. 1937. Some Important Philippine Forestry Problems. *Makiling Echo*, 16(1), 4.

———. 1949. Utilization of Wastelands and the Economics and Development of Reforestation in the Philippines. *Philippine Journal of Forestry*, 6(4), 199–216.

Daproza, Juan. 1931. Forestry Problems. *Makiling Echo*, 10(1), 2–3.

de Beer, Jenne H., & McDermott, Melanie H. 1999. *The Economic Value of Non-timber Forest Products in Southeast Asia*. Amsterdam: Netherlands Committee for IUCN.

De la Cadena, Marisol. 2015. *Earth Beings: Ecologies of Practice across Andean Worlds*. Durham, NC: Duke University Press.

Denison, Wilford. 1915. *Report of the Philippine Commission to the Secretary of War—July 1, 1913, to December 31, 1914*. Washington, DC: Government Printing Office.

Douglas, Mary. 1992. *Risk and Blame: Essays in Cultural Theory*. London: Routledge.

Dove, Michael R. 1983. Theories of Swidden Agriculture, and the Political Economy of Ignorance. *Agroforestry Systems*, 1(2), 85–99.

———. 1997. The Epistemology of Southeast Asia's Anthropogenic Grasslands: Issues of Myth, Science and Development. *Japanese Journal of Southeast Asian Studies*, 35(2), 223–39.

———. 2000. The Life-Cycle of Indigenous Knowledge and the Case of Natural Rubber Production. In R. Ellen, P. Parkes, & A. Bicker (Eds.), *Indigenous Environmental Knowledge and Its Transformations* (pp. 213–51). London: Routledge.

Dressler, Wolfram. 2009. *Old Thoughts in New Ideas: State Conservation Measures, Development and Livelihood on Palawan Island*. Quezon City: Ateneo de Manila University Press.

Dressler, Wolfram, & Pulhin, Juan. 2010. The Shifting Ground of Swidden Agriculture on Palawan Island, the Philippines. *Agriculture and Human Values*, 27(4), 445–59.

Dwyer, Peter, & Minnegal, Monica. 2000. El Niño, Y2K and the "Short, Fat Lady": Drought and Agency in a Lowland Papua New Guinean Community. *Journal of the Polynesian Society*, 109(3), 251–72.

Eder, James F. 1987. *On the Road to Tribal Extinction: Depopulation, Deculturation, and Adaptive Well-Being among the Batak of the Philippines*. Berkeley: University of California Press.

———. 1999. *A Generation Later: Household Strategies and Economic Change in the Rural Philippines*. Honolulu: University of Hawaii Press.

———. 2006. Land Use and Economic Change in the Post-frontier Upland Philippines. *Land Degradation & Development*, 17(2), 149–58.

———. 2010. Muslim Palawan Diversity and Difference on the Periphery of Philippine Islam. *Philippine Studies*, 58(3), 407–20.

Eder, James F., & Fernandez, Janet O. 1996. Palawan, a Last Frontier. In J. F. Eder & J. O. Fernandez (Eds.), *Palawan at the Crossroads* (pp. 1–22). Quezon City: Ateneo de Manila University Press.

Elden, Stuart. 2013. *The Birth of Territory*. Chicago: University of Chicago Press.

Elliot, Charles Burke. 1917. *The Philippines to the End of the Commission Government*. Indianapolis: Bobbs-Merrill.

Ellis, Frank. 2000. *Rural Livelihood Diversity in Developing Countries*. Oxford: Oxford University Press.

Elvin, Mark. 1998. Who Was Responsible for the Weather? Moral Meteorology in Late Imperial China. *Osiris*, 13(1), 213–37.

Escobar, Arturo. 2008. *Territories of Difference: Place, Movements, Life, Redes*. Durham, NC: Duke University Press.

Executive Order 263. 1995. *Adopting Community-Based Forest Management as the National Strategy to Ensure the Sustainable Development of the Country's Forestlands Resources and Providing Mechanisms for Its Implementation*. Manila: Malacañang.

FAO. 1957. Shifting Cultivation. *Tropical Agriculture*, 34(3), 159–64.

———. 1984. *Improved Production Systems as an Alternative to Shifting Cultivation*. Rome: Food and Agriculture Organisation.

Florece, Leonardo M., Lubag, P. O., & Viegas, D. X. 2002. *Quantifying Fire Intensity and Cost of Damage of Forest Fire in Palawan, Philippines*. Paper presented at the Wildland Fire Safety Summit, 4th International Conference on Forest Fire Research, Luso, Portugal.

Ford, James, Maillet, Michelle, Pouliot, Vincent, Meredith, Thomas, Cavanaugh, Alicia, & IHACC Research Team. 2016. Adaptation and Indigenous

Peoples in the United Nations Framework Convention on Climate Change. *Climatic Change, 139*(3), 429–43.
Forsyth, Tim, & Walker, Andrew. 2008. *Forest Guardians, Forest Destroyers: The Politics of Environmental Knowledge in Northern Thailand.* Seattle: University of Washington Press.
Foucault, Michel. 1978. *The History of Sexuality: An Introduction.* Vol. 1. New York: Vintage.
Fowler, Cynthia. 2012. *Ignition Stories: Indigenous Fire Ecology in the Indo-Australian Monsoon Zone.* Durham, NC: Carolina Academic Press.
Fox, Robert. 1954. *Tagbanua Religion and Society.* PhD thesis, University of Chicago.
Freeman, Derek. 1955. *Iban Agriculture: A Report on the Shifting Cultivation of Hill Rice by the Iban of Sarawak.* London: HM Stationery Office.
Fujisaka, Sam., & Capistrano, Doris. 1986. Upland Development in Calminoe: The Roles of Resource Use, Social System and National Policy. In S. Fujisaka, P. E. Sajise, & R. del Castillo (Eds.), *Man, Agriculture and the Tropical Forest: Change and Development in the Philippine Uplands* (pp. 223–44). Bangkok: Winrock International Institute for Agricultural Development.
Gagné, Karine. 2019. *Caring for Glaciers: Land, Animals, and Humanity in the Himalayas.* Seattle: University of Washington Press.
Gauld, Richard. 2000. Maintaining Centralized Control in Community-Based Forestry: Policy Construction in the Philippines. *Development and Change, 31*(1), 229–54.
Gibson, Thomas. 1986. *Sacrifice and Sharing in the Philippine Highlands: Religion and Society among the Buid of Mindoro.* London: Athlone.
Giddens, Anthony. 1999. Risk and Responsibility. *Modern Law Review, 62*(1), 1–10.
Gonzalez, Nestor V. M. 1954. *Children of the Ash-Covered Loam and Other Stories.* Manila: Benipayo Press.
Grove, Kevin. 2013. Biopolitics. In C. Death (Ed.), *Critical Environmental Politics* (pp. 22–30). London: Routledge.
———. 2018. *Resilience.* New York: Routledge.
Gülcur, Macid Y. 1965. *Renewable Natural Resources and Their Problems in the Philippines.* Paper presented at the Conference on Conservation of Nature and Natural Resources in Tropical South East Asia, Bangkok, Thailand.
Heatherington, Tracey. 2010. *Wild Sardinia: Indigeneity and the Global Dreamtimes of Environmentalism.* Seattle: University of Washington Press.
Henley, David. 2005. *Fertility, Food and Fever: Population, Economy and Environment in North and Central Sulawesi, 1600–1930.* Leiden: KITLV Press.

Hilario, Flaviana D., de Guzman, R., Ortega, D., Hayman, P., & Alexander, B. 2009. El Niño Southern Oscillation in the Philippines: Impacts, Forecasts, and Risk Management. *Philippine Journal of Development*, 36(1), 10–34.

Hobbes, Marieke. 2000. *Pala'wan Managing Their Forest: The Case of the Pala'wan of Saray on Southern Palawan Island, the Philippines.* Master's thesis, Leiden University.

Hughes, David McDermott. 2013. Climate Change and the Victim Slot: From Oil to Innocence. *American Anthropologist*, 115(4), 570–81.

———. 2017. *Energy without Conscience: Oil, Climate Change, and Complicity.* Durham, NC: Duke University Press.

Hulme, Mike. 2017. *Weathered: Cultures of Climate.* London: SAGE.

Hunting Technical Services Limited. 1985. *Strategic Environmental Plan for Mainland Palawan.* Boreham: Hunting Technical Services Limited.

Ibrahim, Hindou Oumarou. 2018. Without Traditional Knowledge, There Is No Climate Change Solution. Conservation International. Retrieved from: www.conservation.org/blog/without-traditional-knowledge-there-is-no-climate-change-solution.

Ingold, Tim. 2000. *The Perception of the Environment: Essays in Livelihood, Dwelling and Skill.* London: Routledge.

Internews. 2018. Indigenous Peoples Suffer from the Most Adverse Impacts of Climate Change, Says Victoria Tauli-Corpuz. Internews. Retrieved from: www.internews.org/news/indigenous-peoples-suffer-most-adverse-impacts-climate-change-says-victoria-tauli-corpuz.

Jasanoff, Sheila. 2010. A New Climate for Society. *Theory, Culture & Society*, 27(2–3), 233–53.

Jordana y Morera, Ramon. 1876. *Memoria sobre la producción de los montes públicos de Filipinas durante el año económico de 1873–74.* Madrid: Establecimientos Tipográficos de Manuel Minuesa.

Keesing, Felix Maxwell. 1962. *The Ethnohistory of Northern Luzon.* Stanford: Stanford University Press.

Krupnik, Igor, & Jolly, Dyanna, Eds. 2001. *The Earth Is Faster Now: Indigenous Observations of Arctic Environmental Change.* Fairbanks: Arctic Research Consortium of the United States.

Kummer, David. 1992. *Deforestation in the Postwar Philippines.* Chicago: University of Chicago Press.

Kwiatkowski, Lyn M. 1998. *Struggling with Development: The Politics of Hunger and Gender in the Philippines.* Boulder: Westview.

Lata, Shalini, & Nunn, Patrick. 2012. Misperceptions of Climate-Change Risk as Barriers to Climate-Change Adaptation: A Case Study from the Rewa Delta, Fiji. *Climatic Change*, 110(1), 169–86.

Leach, Edmund. 1954. *Political Systems of Highland Burma*. Cambridge: Harvard University Press.

Li, Tania M. 2007. *The Will to Improve*. Durham, NC: Duke University Press.

———. 2014. *Land's End: Capitalist Relations on an Indigenous Frontier*. Durham, NC: Duke University Press.

Lindayati, Rita. 2002. Ideas and Institutions in Social Forestry Policy. In C. J. P. Colfer & P. Resosudarmo (Eds.), *Which Way Forward? People, Forests, and Policymaking in Indonesia* (pp. 36–39). Bogor: Center for International Forestry Research.

Livingstone, David N. 1999. Tropical Climate and Moral Hygiene: The Anatomy of a Victorian Debate. *British Journal for the History of Science*, 32(1), 93–110.

Lopez, Maria E. Z. 1986. *The Palaw'an: Land, Ethnic Relations and Political Process in a Philippine Frontier System*. PhD thesis, Harvard University.

Lopez, Rogelio M. 1968. *Agricultural Practices of the Manobo in the Interior of Southwestern Cotabato (Mindanao)*. Cebu City: University of San Carlos.

Lynch, Owen J. 1986. Philippine Law and Upland Tenure. In S. Fujisaka, P. E. Sajise, & R. del Castillo (Eds.), *Man, Agriculture and the Tropical Forest: Change and Development in the Philippine Uplands* (pp. 269–92). Bangkok: Winrock International Institute for Agricultural Development.

Lyon, Bradfield, Cristi, Hannagrace, Verceles, Ernesto R., Hilario, Flaviana D., & Abastillas, Rusy. 2006. Seasonal Reversal of the ENSO Rainfall Signal in the Philippines. *Geophysical Research Letters*, 33(24).

Macdonald, Charles. 1992. Invoking the Spirits in Palawan: Ethnography and Pragmatics. In K. Bolton & H. Kwok (Eds.), *Sociolinguistics Today* (pp. 244–60). London: Routledge.

———. 1993. Protestant Missionaries and Palawan Natives: Dialogue, Conflict or Misunderstanding? *Journal of the Anthropological Society of Oxford*, 23(2), 127–37.

———. 1997. Cleansing the Earth: The "Pänggaris" Ceremony in Palawan. *Philippine Studies*, 45(3), 408–22.

———. 2007. *Uncultural Behaviour: An Anthropological Investigation of Suicide in the Southern Philippines*. Honolulu: University of Hawaii Press.

———. 2011. Kinship and Fellowship among the Palawan. In T. Gibson & K. Silander (Eds.), *Anarchic Solidarity: Autonomy, Equality and Fellowship in Southeast Asia* (pp. 119–40). New Haven: Yale University Press.

———. 2013. The Filipino as Libertarian: Contemporary Implications of Anarchism. *Philippine Studies: Historical and Ethnographic Viewpoints*, 61(4), 413–36.

Magno, Francisco. 2001. Forest Devolution and Social Capital: State–Civil Society Relations in the Philippines. *Environmental History*, 6(2), 264–86.

Makil, Perla Q. 1984. Forest Management and Use: Philippine Policies in the Seventies and Beyond. *Philippine Studies*, 32(1), 27–53.

McDermott, Melanie H. 2000. *Boundaries and Pathways: Indigenous Identity, Ancestral Domain, and Forest Use in Palawan, the Philippines*. PhD thesis, University of California, Berkeley.

McElwee, Pamela D. 2016. *Forests Are Gold: Trees, People, and Environmental Rule in Vietnam*. Seattle: University of Washington Press.

Merrill, Elmer D. 1908. *The Forests of Mindanao*. Manila: Bureau of Printing.

———. 1912. Notes on the Flora of Manila with Special Reference to the Introduced Element. *Philippine Journal of Science*, 7(3), 145–208.

Miller, Peter, & Rose, Nikolas. 1990. Governing Economic Life. *Economy and Society*, 19(1), 1–31.

Molintas, Jose M. 2004. Philippine Indigenous Peoples' Struggle for Land and Life: Challenging Legal Texts. *Arizona Journal of International and Comparative Law*, 21, 269–306.

Moore, Donald S. 2005. *Suffering for Territory: Race, Place, and Power in Zimbabwe*. Durham, NC: Duke University Press.

Mydans, Seth. 1997. Southeast Asia Chokes Indonesia's Forest Fires. *New York Times*, September 25, 1997, 1.

Nadasdy, Paul. 2005. Transcending the Debate over the Ecologically Noble Indian: Indigenous Peoples and Environmentalism. *Ethnohistory*, 52(2), 291–331.

Nano, Jose F. 1939. Kaingin Laws and Penalties in the Philippines. *Philippine Journal of Forestry*, 2(2), 87–92.

———. 1951. A Brief History of Philippine Forestry. *Philippine Journal of Forestry*, 8(1–4), 9–128.

Nazarea, Virginia D. 1999. A View from a Point: Ethnoecology as Situated Knowledge. In V. D. Nazarea (Ed.), *Ethnoecology: Situated Knowledge/Located Lives* (pp. 4–20). Tuscon: University of Arizona Press.

Nixon, Robert. 2011. *Slow Violence and the Environmentalism of the Poor*. Cambridge: Harvard University Press.

Novellino, Dario. 2002. The Relevance of Myths and Worldviews in Pälawan Classification, Perceptions and Management of Honey Bees. In J. R. Stepp, F. S. Wyndham, & R. K. Zarger (Eds.), *Ethnobiology and Biocultural Diversity: Proceedings of the 7th International Congress of Ethnobiology* (pp. 189–206). Athens: University of Georgia Press.

———. 2007. Talking About Kultura and Signing Contracts: The Bureaucratization of the Environment on Palawan Island (the Philippines). In C. A. Maida (Ed.), *Sustainability and Communities of Place* (pp. 82–105). New York: Berghahn.

Nygren, Anja. 1999. Local Knowledge in the Environment-Development Discourse: From Dichotomies to Situated Knowledges. *Critique of Anthropology, 19*(3), 267–88.

O'Brien, William E. 2002. The Nature of Shifting Cultivation: Stories of Harmony, Degradation, and Redemption. *Human Ecology, 30*(4), 483–502.

Ocampo, Nilo S. 1996. A History of Palawan. In J. F. Eder & E. Fernandez (Eds.), *Palawan at the Crossroads* (pp. 23–27). Quezon City: Ateneo de Manila University Press.

Olofson, Harold. 1980. Swidden and Kaingin among the Southern Tagalog: A Problem in Philippine Upland Ethno-Agriculture. *Philippine Quarterly of Culture and Society, 8*(2/3), 168–80.

———. 1995. Taboo and Environment, Cebuano and Tagbanuwa: Two Cases of Indigenous Management of Natural Resources in the Philippines and Their Relation to Religion. *Philippine Quarterly of Culture and Society, 23*(1), 20–34.

Orlove, Benjamin S., Chiang, John C. H., & Cane, Mark A. 2000. Forecasting Andean Rainfall and Crop Yield from the Influence of El Niño on Pleiades Visibility. *Nature, 403*, 68.

PCSD Resolution No. 04-233. 2004. *A Resolution Allowing Almaciga Tapping by Indigenous Peoples (IPs) in the Core Zone in Areas Classified by the Palawan Council for Sustainable Development (PCSD) as Tribal Ancestral Lands Pursuant to Section 11 of Republic Act no. 7611, Otherwise Known as the Strategic Environmental Plan (SEP) for Palawan Act.* Puerto Princesa, Palawan Island, the Philippines.

Peluso, Nancy Lee. 1992. *Rich Forests, Poor People: Resource Control and Resistance in Java.* Berkeley: University of California Press.

Peluso, Nancy Lee, & Vandergeest, Peter. 2001. Genealogies of the Political Forest and Customary Rights in Indonesia, Malaysia, and Thailand. *Journal of Asian Studies, 60*(3), 761–812.

Perreñas, Juno. 2018. *Decolonizing Extinction: The Work of Care in Orangutan Rehabilitation.* Durham, NC: Duke University Press.

Pflueger, Owen W. 1930. The "Kaingin" Problem in the Philippines and a Possible Method of Control. *Journal of Forestry, 28*(1), 66–71.

Philippines REDD-Plus Strategy Team. 2010. *The Philippine National Redd-Plus Strategy.* Retrieved from: www.elaw.org/system/files/PhilippineNationalREDDplusStrategy.pdf.

Population Center Foundation. 1980. *Kaingineros: The Philippine Boat People.* Makati: Population Center Foundation.

Potter, Leslie. 2003. Forests versus Agriculture: Colonial Forest Services, Environmental Ideas and the Regulation of Land-Use Change in Southeast

Asia. In L. Tuck-Po, de Jong, W., & Ken-ichi, A. (Eds.), *Political Ecology of Tropical Forests in Southeast Asia* (pp. 29–71). Kyoto: Kyoto University Press.

Povinelli, Elizabeth A. 2002. *The Cunning of Recognition: Indigenous Alterities and the Making of Australian Multiculturalism*. Durham, NC: Duke University Press.

———. 2016. *Geontologies: A Requiem to Late Liberalism*. Durham, NC: Duke University Press.

Powell, Dana. 2018. *Landscapes of Power: Politics of Energy in the Navajo Nation*. Durham, NC: Duke University Press.

PTFPP. 1997. *Kalikasan at kabuhayan: Bakit at paano mo pamamahalaan ang bukid at kalikasan?* Puerto Princesa: Palawan Council for Sustainable Development.

———. 2002a. *Inogbong Catchment Management Plan*. Puerto Princesa: Palawan Council for Sustainable Development.

———. 2002b. *Pamamaraan sa pangangasiwa ng Inogbong catchment, Bataraza*. Puerto Princesa: Palawan Council for Sustainable Development.

Rafael, Vicente L. 1993. *Contracting Colonialism: Translation and Christian Conversion in Tagalog Society under Early Spanish Rule*. Durham, NC: Duke University Press.

Raintree, John B. 1978. *Extension Research and Development in Malandi: Field Test of a Community-Based Paradigm for Appropriate Technology Innovation among the Tagbanwa of Palawan*. PhD thesis, University of Hawaii, Honolulu.

Republic Act 8371. 1997. *An Act to Recognize, Protect and Promote the Rights of Indigenous Cultural Communities/Indigenous Peoples, Creating a National Commission on Indigenous Peoples, Establishing Implementing Mechanisms, Appropriating Funds Therefor, and for Other Purposes*. Manila: Republic of the Philippines.

Revel, Nicole. 1990. *Fleurs de paroles: Histoire naturelle Palawan*. Paris: Peeters.

———. 1998. The Present Day Importance of Oral Traditions: Their Preservation, Publication and Indexing (with Examples from South-East Asia). In W. Heissig (Ed.), *The Present-Day Importance of Oral Traditions* (pp. 195–206). Weisbaden: Verlag fur Sozialwissenschaften.

Ribot, Jesse. 2014. Cause and Response: Vulnerability and Climate in the Anthropocene. *Journal of Peasant Studies*, 41(5), 667–705.

Roth, D. M. 1983. Philippine Forest and Forestry: 1565–1920. In R. P. Tucker & J. R. Richards (Eds.), *Global Deforestation and the Nineteenth Century World Economy* (pp. 30–49). Durham: Duke University Press.

Rudiak-Gould, Peter. 2013. *Climate Change and Tradition in a Small Island State: The Rising Tide*. London: Routledge.

Sajise, Percy E. 1998. Forest Policy in the Philippines: A Winding Trail Towards Participatory Sustainable Development. In *A Step Toward Forest Conservation Strategy: Current Status on Forests in the Asia-Pacific Region* (pp. 222–34). Tokyo: Institute for Global Environmental Strategies.

Sandalo, Ricardo. 1994. Community-Based Coastal Resource Management: The Palawan Experience. In R. S. Pomeroy (Ed.), *Community Management and Common Property of Coastal Fisheries in Asia and the Pacific: Concepts, Methods and Experiences* (pp. 165–87). Manila: International Center for Living Aquatic Resources.

Sawyer, Frederic H. 1900. *Inhabitants of the Philippines*. London: Sampson, Low and Marston.

Schirmer, Daniel B. 1975. The Philippines Conception and Gestation of a Neo-Colony. *Journal of Contemporary Asia*, 5(1), 53–69.

Scott, Geoffery A. 1979. The Evolution of the Socio-economic Approach to Forest Occupancy (Kaingin) Management in the Philippines. *Philippine Geographical Journal*, 23(2), 58–73.

Scott, James C. 2009. *The Art of Not Being Governed: An Anarchist History of Upland Southeast Asia*. New Haven: Yale University Press.

Scott, William H. 1974. *The Discovery of the Igorots: Spanish Contacts with the Pagans of Northern Luzon*. Manila: New Day.

Sherfesee, William Forsythe. 1915. *Annual Report of the Director of Forestry of the Philippine Islands for the Period July 1, 1913 to June 30, 1914*. Manila: Bureau of Printing.

———. 1916. *Annual Report of the Director of Forestry of the Philippine Islands for the Fiscal Year Ended December 31, 1915*. Manila: Bureau of Printing.

Sheridan, Michael. 2016. Tanzanian Farmers' Discourse on Climate and Political Disorder. In S. A. Crate & M. Nuttall (Eds.), *Anthropology and Climate Change: From Actions to Transformations* (pp. 228–40). New York: Routledge.

Shively, Gerald. 2001. Agricultural Change, Rural Labor Markets and Forest Clearing: An Illustrative Case from the Philippines. *Land Economics*, 77(2), 268–84.

Sigaut, François. 1979. Swidden Cultivation in Europe. A Question for Tropical Anthropologists. *Social Science Information*, 18(4–5), 679–94.

Sivaramakrishnan, Kalyanakrishnan. 1999. *Modern Forests: Statemaking and Environmental Change in Colonial Eastern India*. Stanford: Stanford University Press.

Solway, Jacqueline S. 1994. Drought as a Revelatory Crisis: An Exploration of Shifting Entitlements and Hierarchies in the Kalahari, Botswana. *Development and Change*, 25(3), 471–95.

Sorbara, Lucas. 1998. *Conservation, Development and the Making of the Upland Environment: A Philippine Case Study.* Master's thesis, Dalhousie University.

Spencer, Joseph E. 1966. *Shifting Cultivation in Southeastern Asia.* Berkeley: University of California Press.

Survival International. 2009. *The Most Inconvenient Truth of All: Climate Change and Indigenous Peoples.* London: Survival International.

Tamesis, Florencio. 1947. Forestry Problems of the Philippines. *Philippine Journal of Forestry*, 5(1), 3–6.

Theriault, Noah. 2017. A Forest of Dreams: Ontological Multiplicity and the Fantasies of Environmental Government in the Philippines. *Political Geography*, 58, 114–27.

Tsing, Anna L. 1993. *In the Realm of the Diamond Queen: Marginality in an Out-of-the-Way Place.* Princeton, NJ: Princeton University Press.

———. 2005. *Frinction: An Ethnography of Global Connection.* Princeton, NJ: Princeton University Press.

———. 2017. *The Mushroom at the End of the World: On the Possibility of Life in Capitalist Ruins.* Princeton, NJ: Princeton University Press.

Tucker, Richard. 2000. *Insatiable Appetite: The United States and the Ecological Degradation of the Tropical World.* Berkeley: University of California Press.

UNFCCC. 2015. *United Nations Framework Convention on Climate Change 21st Conference of Parties.* Retrieved from: www.cop21paris.org/about/cop21.

Vandergeest, Peter. 2003. Racialization and Citizenship in Thai Forest Politics. *Society & Natural Resources*, 16(1), 19–37.

van Vliet, Nathalie, Mertz, Ole, Heinimann, Andreas, Langanke, Tobias, Pascual, Unai, Schmook, Birgit, Adams, Cristina, Schmidt-Vogel, Dietrich, Messerli, Peter, Liesz, Stephen, Castella, Jean-Christophe, Jørgensen, Lars, Birch-Thomsen, Torben, Hett, Cornelia, Bech-Bruun, Thilde, Ickowitz, Amy, Vu, Kim Chi, Yasuyuki, Kono, Fox, Jefferson, Padoch, Christine, Dressler, Wolfram, & Ziegler, Alan D. 2012. Trends, Drivers and Impacts of Changes in Swidden Cultivation in Tropical Forest-Agriculture Frontiers: A Global Assessment. *Global Environmental Change*, 22(2), 418–29.

Venturello, Manuel H. 1907. Manners and Customs of the Tagbanuas and Other Tribes of Palawan, the Philippines. *Smithsonian Miscellaneous Collection*, 48, 514–58.

Vergara, Benito M. 1995. *Displaying Filipinos: Photography and Colonialism in Early 20th Century Philippines*. Quezon City: University of the Philippines Press.

Vidal y Soler, Sebastian. 1874. *Memoria sobre el ramo de montes en las islas Filipinas*. Madrid: Ariban.

Warner, Katherine. 1979. *Walking on Two Feet: Tagbanwa Adaptation to Philippine Society*. PhD thesis, University of Michigan, Ann Arbor.

Warren, James F. 1985. *The Sulu Zone 1768–1888*. Quezon City: New Day.

———. 2013. *Climate Change and the Impact of Drought on Human Affairs and Human History in the Philippines, 1582 to 2009*. Working paper, Murdoch University.

Watts, Michael. 1983. *Silent Violence: Food, Famine and Peasantry in Northern Nigeria*. Berkeley: University of California Press.

Webb, Sarah. 2016. Domestic "Eco" Tourism and the Production of a Wondrous Nature in the Philippines. In T. Lewis (Ed.), *Green Asia: Ecocultures, Sustainable Lifestyles, and Ethical Consumption* (pp. 81–98). London: Routledge.

West, Paige. 2006. *Conservation Is Our Government Now: The Politics of Ecology in Papua New Guinea*. Durham, NC: Duke University Press.

Whitehead, John. 1893. *The Exploration of Kina Balu*. London: Gurney and Jackson.

Whitford, Harry N. 1911. *The Forests of the Philippines*. Manila: Bureau of Printing.

Worcester, Dean C. 1909. *Report of the Philippine Commission to the Secretary of War*. Washington, DC: Government Printing Office.

———. 1913. *Report of the Philippine Commission to the Secretary of War*. Washington, DC: Government Printing Office.

INDEX

A

academic knowledge, production of, 135–36
Advancing the Development of the Victoria-Anepahan Communities and Ecosystems (ADVANCE-REDD) project, 129–30, 135
agroforestry, 42–43, 69–71, 72–73, 90
Ahern, George, 39
alcohol: incest and, 121, 152n13; rice wine, 51, 59, 125–27
almaciga resin (*bagtik*), 77, 88, 91, 102
ancestor spirits, 54–55, 58, 59
Ancestral Domains Sustainable Development and Protection Plans, 45
Arctic, 141n1
Asgali, Hadjes, 120

B

bagtik (*almaciga* resin), 77, 88, 91, 102
bahag (loincloth), 126–27, 152n17
bananas, 52, 54, 72, 77, 90, 97, 107, 148n13
Bankoff, Greg, 38
barangay, 25, 148n16
barat (wet period), 115–16, 117
Bataraza (municipality): cleavages between uplands and lowlands, 25–27; establishment of, 66; fieldwork in, 27; Inogbong *barangay*, 25; land classification and management, 66–68, 147n8; map, 24*map*; migrants and indigenous dispossession, 65–66, 147n4; Muslim (Moro) lowlanders, 146n1, 147n3; overview, 23–24; Tausug historical dominance, 63–64
Bataraza Bantay Bayan (Bataraza City Guard), 47, 90–91
Belsky, Jill, 150n11
Bessire, Lucas, 112
Beyer, Henry, 146n1
biopolitics, 10
bisaya (non-Muslim Filipino residents of Palawan), 25
blame, for climate change: approach to, 5–6, 12–13; incest and other moral decline, 3–4, 105, 112, 113–14, 117–19, 121–22, 127–28, 151nn5,9; mountains of blame, 6–7, 12, 130–31; resolving issues of, 136–37;

blame, for climate change (*continued*)
  ritual decline and, 125–27; state restrictions on punishment for incestuous relations and, 122–25, 128; Western accounts and indigenous people, 3, 4–5
Bond, David, 112
Brown, Elaine, 147n10
Brown, William, 143n4
brushland, 49, 145n18. *See also* cogon grasslands
bulag (dry period), 115–16, 117
Bulloch, Hannah, 9
Bureau of Forestry (Bureau of Forest Development), 21–22, 37–38, 39–40, 43–44, 67, 143n8. *See also* Department of Environment and Natural Resources
Burma (Myanmar), 15
burning practices: cattle fodder and, 143n5; El Niño drought and, 98–99; firebreaks, 99, 149n5; overview, 49–50; photographs, 14*fig.*, 50*fig.*; seasonal timing, 49, 115, 117, 145n19, 151n4

C

Canopy (Forest Research Institute newsletter), 43
capitalism, 94
carbon trading, 129–30
cash crops, 37, 54, 77, 99, 148n13
cashew, 69, 72, 148n13
cassava, 28, 52, 53–54, 89, 97, 97*fig.*, 102, 105–7, 150n15
Certificates of Ancestral Domain Claim, 22, 44
Certificates of Stewardship contract, 22, 43

Chambers, Robert, 9
Christians and Christianity, 25, 35, 54, 65–66, 145n25
climate: definition, 114; monsoons, 4, 20, 49, 50, 104, 115, 116; Pala'wan understanding of, 115–17, 150n2; rainfall, 19–20, 95–96, 116
climate change: academic knowledge production and, 135–36; approach to, 5–6, 11–12; blame, resolving issues of, 136–37; indigenous knowledge and, 6, 111, 112–13, 132–34, 150n1; indigenous people and, 3, 4–5, 111, 130, 132, 133–35, 141n1; Pala'wan accounts of incest and other moral decline, 3–4, 105, 112, 113–14, 117–19, 121–22, 127–28, 151nn5,9; Pala'wan participation in addressing, 135; from ritual decline, 125–27; from state restrictions on punishment for incestuous relations, 122–25, 128; vulnerability and, 93–95
coconut, 67, 97, 107
cogon grasslands (*cogonales*), 36, 37–38, 49, 143n5
colonialism, 15–16, 21, 34, 35–39, 67
community-based forest management (CBFM), 17–18, 33, 44, 62–63, 68–70. *See also* Palawan Tropical Forestry Protection Programme; social forestry
Condominas, George, 16
Conelly, W. Thomas, 148n17
Conklin, Harold, 17, 45, 132
conservation: and governmentality and biopolitics, 10; indigenous people and, 11; livelihood and, 8–10, 11; place and, 7–8, 62
Conservation International, 134

constellations, 115, 150n3
corn, 52, 96
cosmology, Pala'wan: Christianity and, 145n25; climate and, 116–17; Empu'/ Diyos, 145n23; overview, 54–55; rice origin story, 55–57; spirits, categorization of, 145n22; *sumbang* (incestuous relationships) and, 119; variation within, 145n24. *See also* rituals
credit (*utang*), 23, 79, 149n6. *See also* debt
cropping, 50–52, 145n21. *See also* cash crops

D

Daproza, Juan, 39–40
debt, 23, 66, 142n10, 149n6. *See also* credit (*utang*)
Deleuze, Gilles, 8
Department of Agriculture (DA), 80
Department of Environment and Natural Resources (DENR), 31, 32, 44, 88, 91. *See also* Bureau of Forestry
Dioscorea hispida (*kedut*; tuber), 100
Douglas, Mary, 5
Dove, Michael, 13, 133
dreams, 54–55
Dressler, Wolfram, 142n14
drought. *See* El Niño drought (1997–98)

E

education, 75
Ellen, Roy, 133
Elliot, Charles Burke, 34
El Niño drought (1997–98): approach to, 95, 108–9; attribution to moral decline, 105; crop changes in response to, 105–7; impacts on cash income, 99–100, 104; impacts on swidden agriculture, 97–99; memories of, 96–97, 104–5; moral economy during, 102–3, 108, 150n12; relief rice, 101–2, 149n9; swidden as buffer against disaster and, 107–8, 150n15; wildfires during, 98, 149n4; wild food alternatives, 100–101, 149nn7,8
El Niño–Southern Oscillation (ENSO), 20, 95–96, 116, 149n2
Elvin, Mark, 141n3
Environmentally Critical Areas Network (ECAN), 70, 71, 72, 75, 77, 82–83, 87
environmental policy. *See* conservation; forest management
Escobar, Arturo: *Territories of Difference*, 8
Executive Order 263 (1995), 44

F

fallow use, 18, 53–54, 72, 150n11
Family Approach to Reforestation (1976), 41
fines, 120, 151n11
fires, 32, 36, 37–38, 98, 149n4. *See also* burning practices
Fischer, Arthur, 39
Food and Agriculture Organization (FAO), 17, 18, 42
Ford, James, 152n1
Forest Act (1904), 67
forest management: against swidden agriculture, agroforestry, 42–43, 69–71, 72–73, 90; colonial approaches, 15–16, 34, 35–39, 67, 143n3; community-based

forest management (*continued*)
approaches, 17–18, 33, 44, 62–63, 68–70; Filipino foresters, 39–40, 143n8; and governmentality and biopolitics, 10; Philippine national policy, 9–10, 21–22, 40–45, 67, 142n8, 147n7; social forestry, 15–17, 17–18, 32–33, 41–43; uplands and, 12, 16. *See also* Palawan Tropical Forestry Protection Programme
Forest Management Bureau, 44
Forest Occupancy Management Program (1974), 41
Forest Research Institute, 43
Foucault, Michel, 10
4Ps (Pantawid Pamilyang Pilipino Program), 152n18
Fowler, Cynthia, 149n1
Fox, Robert, 142n9
Freeman, Derek, 16
frontier settlement, 21, 63, 68, 109

G

gender, 30, 148n15
German Agency for Technical Cooperation (GTZ), 75
gobyerno (government), 91
Gonzalez, Nestor, 143n2
governmentality, 10
grasslands, cogon (*cogonales*), 36, 37–38, 49, 143n5
Grove, Kevin, 10
Guattari, Félix, 8

H

hamlets, 26, 27
Hanunóo people, 17, 45
Harris, Holly, 133
harvests, 52–53
health, 75
Heatherington, Tracey, 132, 133
honey, 146nn28,31, 152n16
Hulme, Mike, 114

I

Ibrahim, Hindou Oumarou, 134
incestuous relationships. See *sumbang*
indigenous knowledge, 6, 111, 112–13, 130, 132–34, 150n1
indigenous people: approach to, 5–6, 11–12; attitudes towards and construction in Philippines, 31–32, 34–35, 142n15; in central Palawan, 25, 142n14; climate change and, 3, 4–5, 111, 112–13, 130, 132, 133–35, 141n1; marginalization in Palawan, 22; participation on Palawan in market economies, 23, 142n10; vulnerability and, 93–95. *See also* Pala'wan people; swidden agriculture
Indigenous Peoples Rights Act (IPRA [1997]), 44–45
Indonesia, 15, 16, 17
Inogbong *barangay*, 25, 71–72. *See also* Bataraza (municipality); Pala'wan people
Inogbong River, 26, 29, 99, 118*fig.*, 149n3
insects and insecticides, 52, 79
Inspeccion General de Montes, 35
Integrated Environmental Program (IEP), 69–70
Integrated Social Forestry Program (1982), 41
International Council for Research in Agroforestry, 42, 70

J

Jasanoff, Sheila, 136
Jordana y Morera, Ramon, 36

K

kabuhayan (livelihoods), 11. *See also* livelihoods
kaingin and *kaingiñero*, use of terms, 19, 144nn12,13. *See also* swidden agriculture
Kaingin Law (Act No. 274 [1901]), 41
kedut (*Dioscorea hispida;* tuber), 100
Keesing, Felix: *The Ethnohistory of Northern Luzon*, 35
kelang bulag (big heat), 96

L

labor, collective, 51, 145n20
Labut, Burit, 86–89, 90, 91, 102, 149n18
Labut, Narhelin, 86
Labut, Nerma, 86, 87, 89
language, and land reform, 147n9
Leach, Edmund, 16
Lefebvre, Henri, 8
lenggam/taw't geba (people of the forest), 48–49, 48*fig.*, 145n17
Li, Tania, 94
lime, *kalamansi*, 54, 72, 77, 148n13
livelihoods: approach to, 8–10, 11–12; climate and, 114–15; El Niño drought and, 96–97, 104, 105; Pala'wan overview, 27–28; PTFPP impacts on, 70–71, 76–85, 131; state interventions, 42–43, 44–45, 62–63, 131; transition away from swidden, 108–10
Livingstone, David, 141n3

Local Government Code (Republic Act No. 7160), 148n16
logging, illegal, 101–2, 149n10
Lopez, Maria, 144n14, 147n10
lowlands, vs. uplands, 146n2
lutlut (thanksgiving for rice ritual), 53, 58

M

Macdonald, Charles, 27, 150n12, 152n15
magtanim talaga (really planted [a lot]), 105
Makiling Echo (Bureau of Forestry staff publication), 40
mango, 69, 72, 148n13
Marangas River, 98, 149n3
Marcos dictatorship, 41
Marshall Islands, 5
matinding bulag (intense heat), 96
McDermott, Melanie, 142n14
megwulak lang (just plant), 105
Merrill, Elmer, 38, 143n4
Meyreg (*panglima*), 117, 121
millet (*aturay*), 52
monsoons, 4, 20, 49, 50, 104, 115, 116. *See also* rainfall
Moore, Donald, 7–8
moral decline, 105, 151n9. See also *sumbang* (incestuous relationships)
moral economy, 102–3, 108, 150n12
Moro (Muslim) lowlanders, 146n1, 147n3
mountains of blame, 6–7, 12, 130–31
Mount Mantalingahan Protected Landscape (MMPL), 61–62, 91
municipalities, 148n16
Muslim (Moro) lowlanders, 146n1, 147n3
Myanmar (Burma), 15

N

Narrazid, Sapudin, 64
National Conference on the Kaingin Problem (1965), 41
Nazarea, Virginia, 113

O

Olofson, Harold, 144n12
Orlove, Benjamin, 151n5

P

Pacific states, small-island, 141n1
Palawan Council for Sustainable Development (PCSD), 69, 148n16. *See also* Palawan Tropical Forestry Protection Programme
Palawan Integrated Development Project (PIADP), 69–70, 148n11
Palawan Island: ADVANCE-REDD project, 129–30, 135; Bataraza municipality, 23–24; community-based forest management, 68–70; forest management, 66–68, 142n8; frontier settlement, 20–21; indigenous–migrant relations in central Palawan, 25, 142n14; indigenous peoples, marginalization of, 22; indigenous peoples, in market economies, 23, 142n10; Inogbong *barangay*, 25; map, 24*map*; Mount Mantalingahan Protected Landscape, 61–62, 91; Puerto Princesa Subterranean River National Park, 62, 142n14; rainfall, 19–20, 95–96, 116; Tausug historical dominance in south, 63–64; uplands as mountains of blame, 6–7, 12, 130–31. *See also* Bataraza (municipality); El Niño drought (1997–98); Pala'wan people; Palawan Tropical Forestry Protection Programme; Philippines; swidden agriculture
Pala'wan people: approach to, 6–7, 11–12, 28–29; on climate and seasonal changes, 115–17, 150n2; on climate change, 3–4, 105, 112, 113–14, 117–19, 121–22, 127–28, 151nn5,9; fear of imprisonment, 84–85, 148n17; fieldwork and positionality among, 27, 29–30, 111; in Inogbong *barangay*, 25; livelihoods, 11, 27–28; marginalization to uplands, 25–27, 67–68; migrants and, 65–66, 147n4, 148n14; participation in climate change issues, 135; past and, 30; social organization, 27; Tausug historical dominance and, 63–64. *See also* blame, for climate change; cosmology, Pala'wan; El Niño drought (1997–98); Palawan Tropical Forestry Protection Programme; rituals; *sumbang* (incestuous relationships); swidden agriculture
Palawan Tropical Forestry Protection Programme (PTFPP): afterlife of, 76–85; agroforestry approach, 70–71, 72–73, 148n13; approach to, 63, 91–92; case study, Burit and Nerma Labut, 85–86, 86–89, 91, 149n18; case study, Murta and Lina Tanduk, 85–86, 89–91; climate change associated with, 126; community-based forest management context, 69–70; community engagement, 73–74;

establishment, 69; on firebreaks, 149n5; friction with indigenous people, 76; governance networks, 79–80; in Inogbong *barangay*, 71–72; objectives, 70–71; perceived decline in rice yields and, 77–78, 78*fig.*; reorganization of target population's lives, 74–76, 83–85, 131, 148n12; transition away from swidden and, 78–79; uneven effects, lowlands vs. uplands, 80–83, 80*tab.*, 81*fig.*, 82*fig.*, 85

panglima (customary leaders), 27, 73

Pantawid Pamilyang Pilipino Program (4Ps), 152n18

Paris Agreement, 133–34

past, production of, 30

Peluso, Nancy, 13

pengkebiyagan (livelihoods), 11. See also livelihoods

pepper, 148n13

Perreñas, Juno, 11

personhood, 144n14

pests and pesticides, 52, 79

Philippine Daily Inquirer (newspaper), 32, 143n1

Philippine Journal of Forestry, 40

Philippines: colonial history and forest management, 15–16, 34, 35–39, 67, 143n3; environmental concerns, 19; indigenous people, 31–32, 34–35, 142n15; lowlands vs. uplands, 146n2; national forest management, 9–10, 21–22, 40–45, 67, 142n8, 147n7; social programs, 152n18. See also Palawan Island; Pala'wan people; swidden agriculture

place, production of, 7–8, 62

postfrontier. See frontier settlement

Povinelli, Elizabeth, 5, 136

Puerto Princesa Subterranean River National Park, 62, 142n14

R

radio, 74

Rafael, Vicente, 34

rainfall, 19–20, 95–96, 116. See also monsoons

Ramos, Fidel, 44

reciprocity, 51, 102–3, 146n30, 150n12

"Reducing Emissions from Deforestation and Degradation" (REDD) program, 130

Regalian Doctrine, 67, 147n6

Report of the Philippine Commission (1914), 141n6

Revel, Nicole, 150n3, 151n5

Revised Forest Code (Presidential Decree No. 705 (1975)), 68

Revised Kaingin Law (Republic Act No. 3701 (1963)), 41

Ribot, Jesse, 93–94

rice: El Niño drought and, 99–100, 104; origin myth, 55–57; paddy farming, transition to, 65–66, 78–80, 82, 82*fig.*, 83, 85, 86–87, 142n13, 147n5; paddy farming and gender, 148n15; perceived decline, 77–78, 78*tab.*; relief rice, 101–2, 149n9; ritual use of, 55, 58–59, 146n27, 152n16; swidden harvests, 52–53; swidden planting practices, 50–52

rice wine, 51, 59, 125–27

rituals: climate change and decline of, 125–27; frequency of, 59, 146n32; for health, 146n29; honey used in, 146nn28, 31, 152n16; reciprocal nature of, 146n30; rice used in, 55, 58–59, 146n27, 152n16; terminology,

rituals (*continued*)
146n26; thanksgiving for rice (*lutlut*), 53, 58; when planting rice, 51. *See also* cosmology, Pala'wan
Rudiak-Gould, Peter, 5
rural households, 9, 93, 109–10, 141n4

S

Sandar, Peminta, 86
Sarawak, 15
Sawyer, Frederic, 147n3
Scott, James, 14–15
self-blame. *See* blame, for climate change
Sherfesee, William Forsythe, 37, 39
Siebert, Stephen, 150n11
simaya (sacrifice/offering), 55, 125, 146n26. *See also* rituals
site selection and clearing, 46–49, 47*fig.*, 145n16
Sloping Agricultural Land Technology (SALT), 43
slow violence, 94
sobrang init (excessive heat), 96
social forestry, 17–18, 33, 41–43. *See also* community-based forest management; Palawan Tropical Forestry Protection Programme
space, production of, 7–8, 62
Spain, 34, 35–36, 63, 67, 143n3
Spencer, Joseph Earl, 17
spirituality. *See* cosmology, Pala'wan; rituals
sugarcane, 142n11, 145n21
Sulu Sultanate, 63–64
sumbang (incestuous relationships): alcohol and, 121, 152n13; boundaries of, 118, 151n8; climate change and, 3–4, 105, 112, 117–19, 121–22, 127–28, 151n5; climate change and state restrictions on punishment, 122–25, 128; contemporary cases and concerns, 120–21; frequency and anxiety about, 122, 152nn14,15; geographical impact, 152n12; Pala'wan cosmology and, 119; ritual executions and fines for, 120; sensory aspects of, 119, 151n10; significance for Palawan indigenous groups, 151n7
Survival International, 93
sweet potato, 28, 52
swidden agriculture (*kaingin/uma*): American administration against, 15–16, 21, 36–39, 67, 141n6, 143n4; approach to, 18–19, 33–34, 46, 59–60, 131–32; as buffer against disaster, 107–8, 150n15; burning practices, 14*fig.*, 49–50, 50*fig.*, 98–99, 115, 117, 143n5, 145n19, 151n4; climatic change and, 117, 151n6; community-based forest management and, 17–18, 68–70; cropping and maintenance, 50–52, 145n21; cultural difference and, 18; definition and overview, 12–13; El Niño drought and, 97–99, 103, 104; fallow use, 18, 53–54, 72, 150n11; field sizes, 144n15; Filipino prejudice against, 19, 39–40, 45–46, 143n9; forest and land management against, 15–16, 17, 21–22, 32–33, 40–45, 66–68, 142n9, 144nn10,11; harvests, 52–53; as heritage, 143n2; *kaingiñero*, 19; labor exchange, 51, 145n20; origins and precolonial development, 13–15; in Pala'wan cosmology, 54–59; perceived decline, 77–78, 78*tab.*; personhood and, 144n14; PTFPP impacts on, 77, 79–85; rainfall patterns and,

20; scholarship on, 16–17, 45; site selection and clearing, 46–49, 47*fig.*, 145n16; Spanish administration against, 35–36; state perceptions of, 13; terminology, 144nn12,13; transition away from, 22–23, 78–80, 86–87, 108–10, 150nn13,14

T

Tamesis, Florencio, 39, 40, 143n8
Tanduk, Lina, 86, 89
Tanduk, Muku, 121, 152n13
Tanduk, Murta, 86, 89–91
taro, 52
Tauli-Corpuz, Victoria, 112, 134
Tausug people, 63–64, 66, 147n3
taw't geba/lenggam (people of the forest), 48–49, 48*fig.*, 145n17
tenurial mechanisms, 22, 43, 44, 144nn10,11
Thailand, 18
tobacco, 52, 142n11
Tribal Learning Center, 75–76
Tsing, Anna, 135; *In the Realm of the Diamond Queen*, 12
Tucker, Richard, 39

U

uma, 144nn12,13. *See also* swidden agriculture
ungsud (sacrifice/offering), 55, 125, 146n26. *See also* rituals
United Nations: Reducing Emissions from Deforestation and Degradation (REDD) program, 130
United Nations Framework Convention on Climate Change (UNFCCC), 133–34, 152n1

United States of America: against swidden agriculture, 15–16, 21, 36–39, 67, 141n6, 143n4; colonial rule in Philippines, 34; logging industry in Philippines and, 143n6; Tausug people in southern Palawan and, 64, 147n3
uplands, 6, 12, 16, 25–27, 130–31, 146n2
utang (credit), 23, 79, 149n6. *See also* debt

V

Vandergeest, Peter, 13
Venus (planet), 151n5
Victoria-Anepahan Mountain Range: Advancing the Development of the Victoria-Anepahan Communities and Ecosystems (ADVANCE-REDD) project, 129–30, 135
Vidal y Soler, Sebastian, 36
Vietnam, 15
vulnerability, 93–95, 109–10, 132–33. *See also* El Niño drought (1997–98)

W

Warner, Katherine, 144n14
Watts, Michael: *Silent Violence*, 93
weeding, 52
West, Paige, 8, 11
Whitehead, John, 146n2
Whitford, Harry, 38, 39, 143n4
wild foods, 100–101, 149nn7,8
women, 30, 148n15
Worcester, Dean, 37, 64, 143n7, 147n3

Y

yams, 52, 102, 149n7

CULTURE, PLACE, AND NATURE
Studies in Anthropology and Environment

Mountains of Blame: Climate and Culpability in the Philippine Uplands, by Will Smith

Sacred Cows and Chicken Manchurian: The Everyday Politics of Eating Meat in India, by James Staples

Gardens of Gold: Place-Making in Papua New Guinea, by Jamon Alex Halvaksz

Shifting Livelihoods: Gold Mining and Subsistence in the Chocó, Colombia, by Daniel Tubb

Disturbed Forests, Fragmented Memories: Jarai and Other Lives in the Cambodian Highlands, by Jonathan Padwe

The Snow Leopard and the Goat: Politics of Conservation in the Western Himalayas, by Shafqat Hussain

Roses from Kenya: Labor, Environment, and the Global Trade in Cut Flowers, by Megan A. Styles

Working with the Ancestors: Mana and Place in the Marquesas Islands, by Emily C. Donaldson

Living with Oil and Coal: Resource Politics and Militarization in Northeast India, by Dolly Kikon

Caring for Glaciers: Land, Animals, and Humanity in the Himalayas, by Karine Gagné

Organic Sovereignties: Struggles over Farming in an Age of Free Trade, by Guntra A. Aistara

The Nature of Whiteness: Race, Animals, and Nation in Zimbabwe, by Yuka Suzuki

Forests Are Gold: Trees, People, and Environmental Rule in Vietnam, by Pamela D. McElwee

Conjuring Property: Speculation and Environmental Futures in the Brazilian Amazon, by Jeremy M. Campbell

Andean Waterways: Resource Politics in Highland Peru, by Mattias Borg Rasmussen

Puer Tea: Ancient Caravans and Urban Chic, by Jinghong Zhang

Enclosed: Conservation, Cattle, and Commerce among the Q'eqchi' Maya Lowlanders, by Liza Grandia

Forests of Identity: Society, Ethnicity, and Stereotypes in the Congo River Basin, by Stephanie Rupp

Tahiti Beyond the Postcard: Power, Place, and Everyday Life, by Miriam Kahn

Wild Sardinia: Indigeneity and the Global Dreamtimes of Environmentalism, by Tracey Heatherington

Nature Protests: The End of Ecology in Slovakia, by Edward Snajdr

Forest Guardians, Forest Destroyers: The Politics of Environmental Knowledge in Northern Thailand, by Tim Forsyth and Andrew Walker

Being and Place among the Tlingit, by Thomas F. Thornton

Tropics and the Traveling Gaze: India, Landscape, and Science, 1800–1856, by David Arnold

Ecological Nationalisms: Nature, Livelihood, and Identities in South Asia, edited by Gunnel Cederlöf and K. Sivaramakrishnan

From Enslavement to Environmentalism: Politics on a Southern African Frontier, by David McDermott Hughes

Border Landscapes: The Politics of Akha Land Use in China and Thailand, by Janet C. Sturgeon

Property and Politics in Sabah, Malaysia: Native Struggles over Land Rights, by Amity A. Doolittle

The Earth's Blanket: Traditional Teachings for Sustainable Living, by Nancy Turner

The Kuhls of Kangra: Community-Managed Irrigation in the Western Himalaya, by Mark Baker